T0332405

Arctic Sustainability, Key Methodologies and Knowledge Domains

This book provides a first-ever synthesis of sustainability and sustainable development experiences in the Arctic. It presents state-of-the-art thinking about sustainability for the Arctic from a multi-disciplinary perspective.

This book aims to create a comprehensive, integrative knowledge base for the assessment of Arctic sustainability for countries such as the United States, Canada, Greenland, Iceland, Norway, Sweden, Finland, and Russia, alongside emerging ideas about sustainable development in the Arctic. These ideas relate to understanding how a community's geography matters in determining the required sustainability efforts, decolonial thinking for building sustainability that is crafted by and for local and Indigenous communities, and the idea of polycentrism (i.e., that the paths toward sustainability differ among places and communities). This volume also highlights the recent thinking about sustainability and resilience over the past decade for the rapidly changing Arctic region.

With patterns of thinking drawn from economic, social, environmental, community, and other components of sustainability; observations and monitoring; engagement of Indigenous knowledge; and integration with policy and decision making, the book helps us understand the complexity and interconnectedness of current Arctic transformations in a more comprehensive way.

Jessica K. Graybill is Associate Professor in the Department of Geography at Colgate University, Hamilton, New York. Her research interests include resilience studies and socio-ecological transformations in postsocialist, urban, and remote spaces and grappling with how environments, livelihoods, and possible futures are co-created by multiple actors.

Andrey N. Petrov is Associate Professor and Director of the ARCTICenter at the University of Northern Iowa. Dr. Petrov is an economic geographer who specializes in Arctic sustainable development, economic organization, and changing Arctic social-ecological systems. He is the president of the International Arctic Social Sciences Association.

Routledge Research in Polar Regions
Series Editor: Timothy Heleniak
Nordregio International Research Centre, Sweden

The Routledge series in Polar Regions seeks to include research and policy debates about trends and events taking place in two important world regions: the Arctic and Antarctic. Previously neglected periphery regions, with climate change, resource development and shifting geopolitics, these regions are becoming increasingly crucial to happenings outside these regions. At the same time, the economies, societies and natural environments of the Arctic are undergoing rapid change. This series seeks to draw upon fieldwork, satellite observations, archival studies and other research methods which inform about crucial developments in the Polar regions. It is interdisciplinary, drawing on the work from the social sciences and humanities, bringing together cutting-edge research in the Polar regions with the policy implications.

Resources and Sustainable Development in the Arctic
Edited by Chris Southcott, Frances Abele, Dave Natcher, and Brenda Parlee

Performing Arctic Sovereignty
Policy and Visual Narratives
Corine Wood-Donnelly

Resources, Social and Cultural Sustainabilities in the Arctic
Edited by Monica Tennberg, Hanna Lempinen, and Susanna Pirnes

Arctic Sustainability, Key Methodologies and Knowledge Domains
A Synthesis of Knowledge I
Edited by Jessica K. Graybill and Andrey N. Petrov

For more information about this series, please visit: www.routledge.com/Routledge-Research-in-Polar-Regions/book-series/RRPS

Arctic Sustainability, Key Methodologies and Knowledge Domains

A Synthesis of Knowledge I

Edited by Jessica K. Graybill and Andrey N. Petrov

Routledge
Taylor & Francis Group

LONDON AND NEW YORK

First published 2020
by Routledge
2 Park Square, Milton Park, Abingdon, Oxon OX14 4RN

and by Routledge
52 Vanderbilt Avenue, New York, NY 10017

Routledge is an imprint of the Taylor & Francis Group, an informa business

British Library Cataloguing-in-Publication Data
A catalogue record for this book is available from the British Library

Library of Congress Cataloging-in-Publication Data
Names: Graybill, Jessica K., 1973– editor. | Petrov, Andrey N., editor.
Title: Arctic sustainability, key methodologies and knowledge domains : a synthesis of knowledge I/edited by Jessica K. Graybill and Andrey N. Petrov.
Description: Abingdon, Oxon ; New York, NY : Routledge, 2020. | Series: Routledge research in polar regions | Includes bibliographical references and index.
Identifiers: LCCN 2019050728 (print) | LCCN 2019050729 (ebook)
Subjects: LCSH: Sustainability—Arctic regions. | Environmental protection—Arctic regions. | Indigenous peoples—Arctic regions—Social conditions. | Arctic regions—Environmental conditions.
Classification: LCC GE160.A68 A75 2020 (print) | LCC GE160. A68 (ebook) | DDC 338.9/2709113—dc23
LC record available at https://lccn.loc.gov/2019050728
LC ebook record available at https://lccn.loc.gov/2019050729

ISBN: 978-0-367-22819-4 (hbk)
ISBN: 978-0-429-27701-6 (ebk)

Typeset in Times New Roman
by Apex CoVantage, LLC

Contents

Figures

Tables

Preface

This volume is a first part of a two-volume series on Arctic sustainability. Writing this book would not have been possible without the Belmont Forum international research initiative, which provided funding for this multinational research team composed of scholars from seven Arctic countries. This book is one of the products of the Arctic Sustainability: A Synthesis of Knowledge project co-funded by the US National Science Foundation, the Russian Foundation for Basic Research, the Research Council of Norway, and Nordfosk. These volumes have emerged as a multiyear effort by a group of researchers who worked individually and collectively to formulate the foundations of Arctic sustainability science and tackle the most relevant issues in Arctic sustainable development. The book united a diverse team of experts from Canada, Denmark, Greenland, Iceland, Norway, Russia, Sweden, and the United States to develop a framework that highlights the state of current understanding, best practices, and metrics for achieving sustainability in the Arctic. The effort took into account not only the social, demographic, economic, and environmental aspects of resilience in creating this framework but also looked at these across a range of scales using an inclusive process that engages Arctic stakeholders, rights holders, and knowledge holders.

Arctic sustainability and sustainable development are 'hot' topics that have relevance both to the everyday lives of Arctic residents and to the future of the entire planet. The aim of this cross-disciplinary publication is to provide diverse representations of sustainability to discern primary features, trends, and indicators of Arctic sustainability focused on rapidly changing Arctic social-ecological systems. We hope this book will be useful to various audiences: academics, who may find theory-reach material and timely methodological discussions; policymakers, who may become acquainted with state-of-the-art thinking on sustainability and sustainable development issues in the Arctic; and Arctic residents, who may use this knowledge to advance their understanding circumpolar experiences and their own opportunities regarding sustainable development.

The ideas of the two volumes came from the discussions many of the authors started during the Arctic FRontiers Of SusTainability: Resources, Societies, Environments and Development in the Changing North (Arctic-FROST) project. Launched in 2014, this community has become a home for Arctic sustainability scholarship over the past several years. Annual conferences and workshops, as well as Arctic-FROST publications paved the way to developing the synthesis product presented here. It is especially important to highlight that many of the authors of this book have been Arctic-FROST early career fellows or senior scholars who served as their mentors.

Another remarkable characteristic of this book and the subsequent volume is the diversity of the disciplinary knowledge of the authors. We comprise anthropologists, geographers, biologists, ecologists, economists, sociologists, and Indigenous studies experts, among others.

Sustainable development in the Arctic is a collective challenge and collective opportunity. Sustainability is both a process and outcome, and it concerns all Arctic residents and everyone around the world because what happens in the Arctic does not stay in the Arctic. We hope this publication will inspire a new generation of Arctic sustainability scholars, stimulate further research, assist in policymaking, and build dialog with sustainability scholarship and practice in other world regions.

Contributors

Shauna BurnSilver is Associate Professor in the School of Human Evolution and Social Change and Senior Sustainability Scientist at the Julie Ann Wrigley Global Institute of Sustainability, both at Arizona State University. She is an environmental anthropologist who studies how global climate and economic changes affect relationships between hunter-fisher communities in the Alaskan Arctic and the environments they depend on. At the core of her research is an examination of modern mixed economies as a livelihood lens through which households and communities engage with change. A majority of her research is interdisciplinary and collaborative, facilitating examination of broader questions around sustainability, vulnerability, and resilience at the scale of households and communities within social-ecological systems. Dr. BurnSilver's work with communities is process focused, incorporating unique insights gained from social network analysis, local knowledge, collaborative science, and social-ecological modeling to understand patterns of change—and their implications for human wellbeing and ecosystems.

Tatiana Degai holds a PhD from the University of Arizona and is currently a postdoctoral scholar at the ARCTICenter, University of Northern Iowa and Member of the Itelmen Peoples Council "Tskhanom". Dr. Degai is an Indigenous scholar from Kamchatka, Russia who specializes in Indigenous studies, Indigenous languages, knowledge, and culture, language revitalization, sustainable development in the Arctic, Indigenous methodologies, as well as sociolinguistics.

Aileen A. Espíritu is a researcher at the Barents Institute at the University of Tromsø, The Arctic University of Norway. She was previously an Assistant Professor (Tenured) at the University of Northern British Columbia, Canada, teaching Northern Studies, Soviet and Russian social history and politics, indigenous, and gender studies. She has published on the impact of industrialization on indigenous peoples in Siberia and more generally on Circumpolar Northern communities. She has also published on her current research on the comparative study of border identities, border crossings, and life on the borderlands of Europe especially in an expanded EU. Aileen has ongoing research on sustainable development in the Arctic regions, notably its urban areas; region-building in the Arctic and the Barents Region; identity politics in indigenous and non-indigenous Northern communities; the impact of industrialization and post-industrialization on mono-industry towns in the High North; and the politics of community sustainability in Russia in comparative perspective.

Gail Fondahl is Professor of Geography at the University of Northern British Columbia. Her PhD is from the University of California–Berkeley. Professor Fondahl's research focuses on the legal geographies of indigenous territorial rights in the Russian North and the cultural and governance dimensions of Arctic sustainability. She served president of the International Arctic Social Sciences Association (2011–2014) and Canada's representative to the International Arctic Science Committee's Social & Human Sciences Working Group (2011–2018). Dr. Fondahl co-edited the second *Arctic Human Development Report* (2014) and *Northern Sustainabilities: Understanding and Addressing Change in the Circumpolar World* (Springer, 2017).

Susanna Gartler is a PhD candidate at the department of anthropology of the University of Vienna. She is interested in the topics of re-indigenization, the extractive industry, indigenous planning, sustainability, climate change, oral history, and science communication and outreach, as well as socio-cultural aspects of permafrost thaw. As a student investigator in the project LACE—Labour Mobility and Community Participation in the Extractive Industry. Case Study in the Canadian North, she has worked in the Yukon Territory for the past six years. Susanna is also a collaborator in Nunataryuk—"Permafrost thaw and the changing Arctic coast, science for socioeconomic adaptation," in which her focuses on identifying social indicators of climate change, risks associated with permafrost thaw and adaptation and equitable mitigation strategies in the Beaufort Sea Delta. Her doctoral thesis focuses on First Nation of Na-Cho Nyäk Dun oral history and the planning of an indigenous cultural center in the Yukon Territory.

Jessica K. Graybill is Associate Professor in the Department of Geography at Colgate University, Hamilton, New York. Her research interests include resilience studies and socio-ecological transformations in postsocialist, urban, and remote spaces and grappling with how environments, livelihoods, and possible futures are co-created by multiple actors. Her regional focus in on the Eurasian Arctic and Far East and on shrinking cities in other contexts, specifically in multicultural and refugee-repopulated Utica, New York. Currently, she is a co-principal investigator on two National Science Foundation grants (Arctic Frontiers of Resources and Sustainability and Arctic Coast), and her research relates to resource use, extraction, and climate change in the Russian Arctic and Far East. She is the editor of *Polar Geography*, and recent publications include the forthcoming book, *Cities of the World*, 7th edition and "Emotional Environments of Energy Extraction in Russia" (2019) in the *Annals of the American Association of Geographers*.

Klaus Georg Hansen holds a PhD in planning, social anthropology, and Greenlandic culture and language. He currently serves as the head of the Interior Division for the Government of Greenland. His research topics include the colonial history of Greenland, political development in Greenland, kayak dizziness, demography in the Arctic, and large-scale industrial development in the Arctic. He has previously held positions as the head of the national library of Greenland, head of Sisimiut Museum, head of division of national spatial planning, head of faculty at Ilisimatusarfik, and deputy director at Nordregio in Sweden. His recent book publication is *From Passive Observers to Active Participants. Mapping the Mechanisms Behind the Last 150 Years of Social Change and the Gradual Process of Democratization in Greenland* (2019).

Diane Hirshberg is Professor of Education Policy at the Institute of Social and Economic Research, University of Alaska Anchorage. She also is a board member for the Arctic Research Consortium of the U.S. and a council member for the International Arctic Social Sciences Association. Her research interests include education policy analysis, indigenous education, circumpolar education issues, and the role of education in sustainable development. She co-edited the recent book publication *Including the North: A Comparative Study of the Policies on Inclusion and Equity in the Circumpolar North* (2019).

Lee Huskey is Emeritus Professor of Economics at the University of Alaska Anchorage. He has been chair of the Economics Department and worked with the university's Center for Economic Education. Huskey was president of the Western Regional Science Association (WRSA) in 2005 and organized special paper sessions at the WRSA meetings for social scientists working in remote regions for over 30 years. His research has concentrated on the economics of remote regions, in particular the rural regions of Alaska. His recent research on migration in the regions of the Circumpolar North was supported by the US National Science Foundation. Huskey has presented his work in many international forums, co-edited books, and published papers on the special economics of remote economies, migration in Alaska, and the teaching of economics.

Gary Kofinas is Professor Emeritus at the University of Alaska Fairbanks (UAF) with the School of Natural Resources and Institute of Arctic Biology. He is the former co-director of the Resilience and Adaptation graduate program in sustainability at UAF, a lead author of the IPCC Special Report on Oceans and the Cryosphere/Polar Regions chapter, and has served as a principal investigator and researcher on many studies examining the sustainability of rural Arctic communities. His research focuses on the resilience of social-ecological systems, community-based conservation, the use of local and traditional knowledge in monitoring and research, and institutional arrangements for the co-management of natural resources. He received his PhD in Interdisciplinary Studies/Resource Management Science from the University of British Columbia.

Vera Kuklina is Research Professor in the Department of Geography, George Washington University and Senior Research Associate at the V.B. Sochava Institute of Geography of the Siberian Branch of Russian Academy of Sciences. Her research interests include urbanization of indigenous people, traditional land use, socio-ecological systems, cultural geographies of infrastructure, and remoteness. Recent publications include "Power of Rhythms—Trains and Work Along the Baikal-Amur Mainline (BAM) in Siberia" (2019) and "The Roads of the Sayan Mountains: Theorizing Remoteness in Eastern Siberia" (2018). Dr. Kuklina currently leads a project titled "Informal Roads: The Impact of Unofficial Transportation Routes on Remote Arctic Communities," which aims to provide an interdisciplinary analysis of the overall impact of informal roads on Arctic environmenst and economic, social, and cultural wellbeing of local communities.

Joan Nymand Larsen is Professor of Economics and Arctic Studies at the University of Akureyri; Senior Scientist and Research Director at the Stefansson Arctic Institute; and Adjunct Professor in the Department of Economics and Business at the University of Greenland. She specializes in processes of economic, sustainable, and human development in the Arctic; northern extractive industries; and climate change impacts and adaptation. Her field research includes her study of the issues and challenges facing young people in the Arctic; her research on impacts of thawing permafrost along the Arctic coast and strategies for adaptation and mitigation (EU H2020 Nunataryuk); and the study of Arctic communities and resource extraction (REXSAC). Among notable publications are the *Arctic Human Development Report* (2014); *The New Arctic*

(2015); *Arctic Social Indicators* (2010, 2014); *Polar Regions in Impacts, Adaptation, and Vulnerability* (2014); and *Arctic Marine Resource Governance and Development* (2018).

Andrey N. Petrov is Associate Professor of Geography and director of the ARCTICenter at the University of Northern Iowa. Dr. Petrov is an economic geographer who specializes in Arctic economy, regional development, and post-Soviet society, with an emphasis on the social geography of indigenous populations of Russia. His current research is focused on regions of the Russian and Canadian North and concerns sustainable development, spatial organization, and changing Arctic social-ecological systems. Dr. Petrov leads the Research Coordination Networks in Arctic Sustainability (Arctic-FROST) and Arctic Coastal Resilience (Arctic-COAST). Dr. Petrov is the president of the International Arctic Social Sciences Association and chair of the International Arctic Science Committee Social and Human Working Group.

Peter Schweitzer is Professor and Chair at the Department of Social and Cultural Anthropology of the University of Vienna and Professor Emeritus at the University of Alaska Fairbanks. His theoretical interests range from kinship and identity politics to human–environmental interactions, including the social lives of infrastructure and the community effects of global climate change; his regional focus areas include the Circumpolar North and the former Soviet Union. Schweitzer is past president of the International Arctic Social Sciences Association and past chair of the Social and Human Sciences Working Group of the International Arctic Science Committee. He is the editor of *Dividends of Kinship* (Routledge, 2000), as well as co-editor of *Hunters and Gatherers in the Modern World* (Berghahn, 2000), *Arctic Social Indicators* (Nordic Council of Ministers, 2010), and *Arctic Social Indicators II: Implementation* (Nordic Council of Ministers, 2014). Schweitzer is also a co-author of *Russian Old-Settlers of Siberia* (Novoe izdatel'stvo, 2004; in Russian) and of *Arctic Sustainability Research* (Routledge, 2017).

Chris Southcott is Professor of Sociology at Lakehead University and the Principal Investigator for the SSHRC MCRI-sponsored research network Resources and Sustainable Development in the Arctic. Raised in Northern Canada, he has been involved in community-based research in the Circumpolar North for over 30 years. During these years, he has published over 120 scientific reports, books, book chapters, and articles dealing with social and economic change in Northern Canada and the rest of the circumpolar world. He has authored, edited, or co-edited eight books, including *Globalization and the Circumpolar North*

(University of Alaska Press), *Northern Communities Working Together: The Social Economy of Canada's North* (University of Toronto Press), and *Resources and Sustainable Development in the Arctic* (Routledge). Over the past 18 years, he has led several major Canadian and international research initiatives dealing with social and economic development in northern regions.

Tatiana Vlasova is a leading researcher at the Institute of Geography, Russian Academy of Sciences. Her experience in the Arctic is based on the fieldwork and participation in several international multidisciplinary projects such as: Arctic Climate Impact Assessment, Local Health and Environmental Reporting from the Indigenous Peoples of the Russian North (UNEP Grid-Arendal), Arctic Resilience Report. She served as a member of the 2008–09 IPY Committee of Russia and on the Sub-Committee on Observations. She has been a co-principle investigator and the leader of a Russian team of the Belmont forum "Arctic Sustainability: Synthesis of Knowledge" project supported by the Russian Foundation for Basic Research (RFBR). Her current research interests include socially-oriented observations and assessments of quality of life conditions and human capital development involving traditional and local knowledge and Arctic sustainability monitoring. She is the Chair of the International Geographical Union Commission on "Cold and High Altitude Regions" (IGU-CHAR) and Councilor of the International Arctic Social Science Association (IASSA) and a member of the editorial board of *Polar Geography*.

Gary N. Wilson is Professor in the Department of Political Science and the Coordinator of the Northern Studies Program at the University of Northern British Columbia. His research examines politics and governance in the Canadian and Circumpolar North, with a particular focus on Inuit self-government in the Canadian Arctic. He is currently serving as the president of the Association of Canadian Universities for Northern Studies and is a member of the Council of the International Arctic Social Sciences Association.

1 Introduction to Arctic sustainability

A synthesis of knowledge

Jessica K. Graybill and Andrey N. Petrov

Sustainability: a historical idea for the present and future

The big idea

"Sustainable development is development that meet the needs of the present without compromising the ability of future generations to meet their needs." While not the first conceptualization of sustainability, this definition of sustainable development guides scholarship and practitioner use (Brundtland, 1987, p. 41). Whether or not this conceptualization is used or appreciated in its original form now, it provokes scholarly inquiry into the interlocking, ongoing, and expanding crises in global economy and global ecology first acknowledged in the *Limits to Growth* report (1972) and continuing with the United Nation's Sustainable Development Goals. Indeed, over the past three decades, sustainability studies, sustainability science, and practice-oriented sustainability work (in professional fields such as urban or regional planning) have utilized, reformulated, and progressed in ways useful for scholarship and practitioner needs for, in, and with communities while confronting the challenges of the 21st century at local and global scales.

By recognizing the importance of addressing the needs and desires of specific places and of including communities in conceptually framing sustainability, important questions arise about social and environmental justice, equity across communities and generations, and how multiple users (scholars, environmental or social stewards, practitioners) propagate understandings and implementations of the sustainability concept (Kofinas, 2005; Crate, 2006; Graybill, 2009; Fondahl & Wilson, 2017; Petrov et al., 2017; Gad, Jakobsen, & Strandsbjerg, 2017). Critique of the concept (Swyngedouw, 2010) has led scholars, practitioners, and communities to refine what it means to care for people and the Earth, thereby changing ideas about sustainable development and raising questions about what it means to thrive sustainably. We find that the shift in use of this term—from an adjective describing a noun (sustainable development) to a stand-alone noun (sustainability)—allows multiple ideas about sustainability to flourish, an important step forward. Examples of progress are from ecologists understanding sustainability as improving the quality of life while living within the carrying capacity of supporting ecosystems (UNEP, 1991) and environmental justice scholarship calls for "the need to ensure a better quality of life for all now, and into the future, in a just and equitable manner, while living within the limits of supporting ecosystems" (Agyeman, Bullard, & Evans, 2003, p. 5).

As the field of sustainability advances, its conceptualization, professional practices of development, and application progress from local to global scales. Three aspects of advancement in sustainability thinking—place matters, decoloniality, and polycentrism—inform this book.

Place matters

Understanding global sustainability issues requires a place-based perspective. Scholars and practitioners attempting to apply the well-intentioned Brundtland definition of sustainability in real-time settings recognize that sustainability requires "concretization" in place and by and for communities alongside considering how the global(izing) world also impacts communities and places. Attempting to make places, cultures, economies, resource development, or governance structures sustainable requires attention to specific people and regions and understanding how local places interact with global ecologies, economies, and societies. The past two decades indicate a turn toward interdisciplinary examination of socio-ecological systems (SESs), sometimes called coupled natural–human systems, to understand how natural and social systems are intertwined in complex ways (AACA, 2017; ARR, 2013). A centerpiece of sustainability research, an SES perspective integrates disciplines, stakeholders, and knowledge systems in multi-, inter-, and transdisciplinary ways. A systems perspective enhances scholarly understanding of the complexity of human-nature interactions by providing a toolkit to conceptually map the who, what, when, where, and why of these interactions (after Meadows, 2008). This aids scholars and practitioners in examining how socio-ecological issues affect a community (a neighborhood, an ethnic group) or a place (a remote settlement, a resource extraction site). When the system is understood, scholars, practitioners, and community members may better address the root causes of unsustainability and more ably choose paths to sustainability.

Decolonial thinking

Critique of the term 'sustainable development' led to greater evaluation of the idea of development. Attempts to implement the Brundtland conceptualization of sustainable development requires asking by whom and for whom sustainability is proposed, defined, and potentially implemented. Rather than understand such questioning as semantic nitpicking, it is an attempt to decolonize knowledge about, and actions taken in, communities and places that may not be part of the West or the Global North. After all, the Brundtland conceptualization is a product of the Global North. Decoloniality questions the stories and histories of the colonial experience to decenter the powerful "underlying logic of the foundation and unfolding of Western civilization from the Renaissance to today" (Mignolo, 2011, p. 2). A response to the unquestioned domination of people and places by Europeans in political, social, and cultural realms (Quijano, 2007), decoloniality critiques Eurocentric modernity (Denzin, Norman, Lincoln, & Tuhiwai, 2008). Applied to sustainability, decolonial thinking means that local experiences with "glocal" phenomena—such as localized resource extraction for global markets; the expansion of environmental and cultural, especially Indigenous, tourism; or the globalization of culture via media and social media—are part of the life experiences of "border" (i.e., of Eurocentrism, of the Global North; Laurie, 2012) communities. Regarding sustainability, decoloniality provides an analytical toolkit for exploring, defining, and implementing ideas about what is sustainable or not in non-Eurocentric ways.

In the Arctic, decolonial thinking is especially important among Indigenous and remote communities for whom a Eurocentric conceptualization of sustainability may not apply and for whom decolonial efforts are central to 21st-century thriving. Additionally, co-production of knowledge, one of many possible Indigenous methodologies (Bawaka Country, Suchet-Pearson, Wright, Lloyed, & Burrawanga, 2013; Johnson et al., 2016; Lavrillier & Gayshev, 2017), is important for Indigenous peoples to become active members in shaping their communities and defining actions toward potential futures when communicating with scholars and practitioners about problems of and desires for sustainable communities and environments.

Polycentrism

The multiple origins and conceptualizations of sustainability reflect multiple epistemologies, cultures, and places within and beyond the West. For example, Oren Lyons, a former Chief of the Onondaga Nation, eloquently asks us to consider the concept of seven generations sustainability: "We are looking ahead, as is one of the first mandates given us as chiefs, to make sure and to make every decision that we make relate to the welfare and wellbeing of the seventh generation to come." "What about the seventh generation? Where are you taking them? What will they have?" (Lyons, 1980, p. 172). In Russia, Vernadksy's early 20th-century concept of the noösphere as a sphere of human reason (*sfera razuma*) imagined humans as capable of transforming Earth through rational scientific and managerial engagement *with* biogeophysical environments, where humans choose to employ their minds to become agents of evolution. In this way, humans craft a world synergistically with the biosphere to develop societies and technologies that do not destroy the Earth or its human communities (Lavrillier & Gayshev, 2017; Vernadsky, 2002; Pitt & Samson, 1999; Shaw & Oldfield, 2006; Graybill, 2007; Moiseev, 1990). The polycentric origins of the sustainability concept deserve greater attention from scholars and practitioners in the definition, refinement, and application of the concept to actual communities and places.

Foundations of sustainability knowledge

Informed by sustainability scholars, the desires of local—global political actors and, sometimes, the needs of humans and ecosystems, sustainability attempts to provide ethical guiding principles for the wellbeing of SESs. Resilience thinking (Walker & Salt, 2006), applied understandings of vulnerability (Turner et al., 2003; Smit & Wandel, 2006; Haalboom & Natcher, 2012), and adaptation Jacob et al., 2010; Cameron, 2012) underpin ideas about how SESs might be sustained or even flourish. Recognizing SESs as dynamic, in which surprise events punctuate their ever-changing structures and functions, resilience thinking addresses rapid, ongoing change. Resilience—the capacity of a system to absorb disturbance and retain its structure and function—is a system condition and requires understanding where and how a system may be strengthened and when transformation is needed. Resilience is a cornerstone in any plan for sustainable development in our current world characterized by dynamic socio-ecological changes because if a system's resilience is not understood or enhanced, the structure and function of the system may collapse, leading to unsustainable development. Thus, resilience is the key to sustainability when considering the wellbeing or thrivability of humans and ecosystems (Braungart & McDonough, 2002; Oestreicher, et al., 2018). Sustainability research, then, addresses the dynamics of ever-evolving SESs through cross-disciplinary research to "improve the quality of life and endure" (DeVries, 2013, p. 7) future changes.

Closely related to resilience, adaptive capacity in SESs describes how institutions and communities can learn from accumulated knowledge and experience to become flexible in problem solving and governance when faced with ongoing or impending change (Berkes et al., 2003). Knowing how people respond to change is among the least understood aspects of ecosystem management (Holling & Gunderson, 2002) and adaptive capacity research examines four interrelated concerns: living with change and uncertainty, increasing socio-ecological diversity, synthesizing knowledge for community learning, and creating opportunities for communities to become self-organized and thereby more sustainable (Folke et al., 2005). Creating adaptive capacity is a social and political process, so understanding what it is or could be for local communities requires attention to scientific, cultural, and governance concerns.

Systems thinking, resilience theory, and adaptive capacity are critical concepts for addressing environmental, social, economic, and cultural changes (Walker et al., 2006; Chapin, Kofinas, & Folke, 2009; Jacob et al., 2010). Each of these conceptual tools provides a way to analyze the structure and behavior of components in a system and the system as a whole. However, these tools and subsequent analyses of socio-ecological conditions and transformations are often not synthesized, leading to fragmented knowledge of the sustainability potentials of geographically disparate communities. A synthetic approach that combines systems, resilience, and adaptive capacity approaches may allow researchers to understand how systems behave similarly, which "is perhaps our best hope for making lasting changes on any levels" (Meadows, 2008, p. 32). Applying resilience thinking assures that local-global communities (of researchers, policymakers, local communities) interact to address sustainable development. Addressing adaptive capacity requires combining scientific knowledge with social-political action, a necessity for ensuring community. In combination, these tasks require transdisciplinary teams of researchers and community members to determine the adaptability, vulnerability, resilience, and transformability of human–environment systems (Holling & Gunderson, 2002; Clark & Dickson, 2003; Kates, 2011).

While a synthetic practice and science of sustainability is still emerging, governance is also important to the wellbeing of natural and social systems (Parr, 2012; Armitage, Berkes, & Doubleday, 2010; Petrov et al., 2017). For these reasons, sustainability as a scholarly endeavor and as a social movement is necessarily problem based, focused on solving tangible concerns at local or regional scales. Because of the close collaboration of scientists, communities, and policymakers for sustainable governance, especially regarding rapidly transforming places or phenomena, some researchers understand sustainability scholarship to be related to "post-normal" science, in which "facts are uncertain, values in dispute, stakes high and decisions urgent" (Funtowicz & Ravetz, 1991, p. 138). We emphasize the necessity to include multiple stakeholders—policymakers, educators, Indigenous and local communities—in conceptualizing sustainability for a specific locale, such as the Arctic, or for a specific phenomenon, such as economic growth in an extractive resource hub or climate-induced biophysical change.

Synthesizing sustainability thinking for a rapidly transforming Arctic

The Arctic has become synonymous with change as global environmental change transforms the lives and experiences of people in this region, socioeconomically, politically, culturally, and environmentally. Knowing that Arctic climatic changes will be some of the most dramatic worldwide, many scholars focus on documenting the Arctic's physical changes spatially and longitudinally, determining community vulnerabilities and risks due to rapid biogeochemical and socioeconomic transformations, and creating science-policy frameworks for addressing adaptation to changing environmental conditions (Krupnik, Allison, Bell, Cutler, & Hik, 2011).

For example, recent biogeochemical research indicates myriad intertwined environmental challenges related to global environmental change: re-volatilization of persistent organic pollutants as surface temperatures warm (Ma, Hung, Tian, & Kallenborn, 2011), permafrost thaw and its implications for Arctic landscapes (Raynolds et al., 2014), Arctic greening and ecosystem restructuring (Macias-Faura et al., 2012), and the effects of increased shrub coverage on Arctic albedo (Loranty & Goetz, 2012). Recent social science and humanistic research about Arctic SESs include investigations of vulnerability (and resilience) to anthropogenic climate change, focusing on loss of the ability to pursue traditional livelihoods, threats to ecosystems sustaining human communities and the need to adapt to new environmental regimes (Wilson, 2003; Dinero, 2011; Hovelsrud & Smit, 2010). Much social science research has been undertaken in the western Arctic, most notably in the Canadian North (Ford & Pearce, 2010; Laidler et al., 2009) and Alaska (Chapin, Sommerkorn, Robards, & Hillmer-Pegram, 2015; Kofinas, 2005; BurnSilver, Boone, Kofinas, & Brinkman, 2017; Berman, Kofinas, BurnSilver, Fondahl, & Wilson, 2017). Fewer studies address SESs in non-Western Arctic and sub-Arctic places (especially Russia, Greenland, or Fenno-Scandia), but the number of such studies is increasing (see Voinov et al., 2004; Crate, 2006; Forbes & Stammler, 2009; Forbes, 2013; Keskitalo, Dannevig, Hovelsrud, West, & Swartling, 2011; Graybill, 2013; Vlasova & Volkov, 2016).

Across the Arctic, projected changes in natural systems will impact human systems and may have direct and immediate implications for land use, economies, subsistence livelihoods, and cultures. Ongoing, often rapid transformation caused by biogeochemical, socioeconomic, and political changes raises interest in Arctic natural resources and development by (trans)national actors (Emmerson, 2010; Heininen & Southcott, 2010; Smith, 2011). These changes re-open long-standing questions about the rational use of resources, methods, and goals of ecosystem conservation and paths to greater human wellbeing in this socioeconomically and politically varied region. While some climatic changes in the Arctic may be economically beneficial, such as decreased climate severity and associated heating costs (Hovelsrud, Poppel, Van Oort, & Reist, 2011) or lengthened navigation season (Lindstad, Bright, & Strømman, 2016), many changes may impact the natural environment and traditional and non-traditional economic sectors adversely (ACIA, 2004; Ford, McDowell, & Pearce, 2015; AACA, 2017). Despite recent advancement of Arctic research, understanding of the complex structures, functions, and interactions within or among SESs across this region remains inchoate and will require updating as climatic and socioeconomic transformation continues.

What will be the impact of rapid and ongoing socio-ecological changes on people and places in the Arctic? What choices will Arctic communities have about how and where to live or pursue their livelihoods in the future? How will communities and places make thoughtful, long-term decisions about how or where to thrive? What are the roles of government and scholars in supporting communities? These are only some of the questions that arise in a first discussion of Arctic sustainability.

At this juncture in the research and practice of sustainability in the Arctic, a synthesis of the state of sustainability knowledge about this region is necessary. The sustainability concept has such a strong hold on global public imaginations that it is a common social, and thus political, discourse. Indeed, Parr (2012) writes "popular culture is the predominant arena where the meaning and value of sustainability is contested, produced and exercised" (p. 3). In the Arctic, sustainability as a way of knowing and as a practice has ever-deepening roots. Understanding how sustainability is used in scientific research and enacted in socio-environmental practices, including policy development and community use, can only assist in understanding how a rapidly transforming Arctic will be addressed from local to global scales.

This book provides a synthesis, based on the issues raised in this chapter, about the state of sustainability for the Arctic region. The synthesis is based on comprehensive review of the state of sustainability research being conducted in the Arctic, noting key trends, concepts, and modes of approaching sustainability studies. The goal of this book is twofold. First, we aim to synthesize existing knowledge about Arctic sustainability to understand the relative importance of, and interactions between, natural and human systems across the North. Second, we aim to create a comprehensive, integrative knowledge base for, and an assessment of Arctic sustainability that is place specific, decolonial, and polycentric. The editors and authors of this volume have worked closely to create a synthesis of knowledge about sustainability in the Arctic that will be useful for scholars, practitioners, and community members, in the hopes that Arctic communities and places may thrive now and into the future.

Operationalizing sustainability in the Arctic region

The Arctic region is home to multiple people living in multiple ways: Indigenous and multiethnic communities reside in well-connected large cities, in smaller settlements related to traditional lifestyles and/or industrial resource extraction, and in more remote (semi)subsistence or (semi)nomadic ways. While transformation has long characterized places and cultures in Arctic history, the rapidity and scale of current changes—climatic and social—imperil present and future communities and places in unprecedented ways. Operationalizing any concept of sustainability in, for, or of the Arctic requires attention to the development of this concept and how usage may vary.

History of sustainability research in the Arctic

The history of conceptualizing and operationalizing sustainability in the Arctic somewhat mirrors processes in other world regions, as attempts to care for the environment have occurred alongside economic growth and societal development (Petrov et al., 2017). However, three additional concepts shape Arctic sustainability. First, the development of Arctic sustainability emphasizes the wellbeing of Indigenous communities, and measures toward this goal are due to local initiatives and the ongoing work of the Arctic Council (Tennberg, 1998; Keskitalo, 2004; Heininen & Plouffe, 2016). Second, "sustainability and sustainable development are inextricably linked to resource exploitation" (Petrov et al., 2017, p. 7). Multiple nations colonized the Arctic as a place from which to extract resources—first whale oil and furs; later hydrocarbons and minerals—which makes extraction the basis of the Arctic economy and the basis of entire settlements and transportation hubs. Third, in October 1987, Mikhail Gorbachev, then leader of the Soviet Union, issued policy initiatives to end the Cold War in the Arctic (Åtland, 2011), creating a "zone of peace which would dissolve "[b]arriers that had precluded efforts to create co-operative arrangements encompassing the Arctic as a distinct region" (Young, 2010, p. 168).

More recently, sustainable developed was only partially addressed until the Arctic Climate Impact Assessment (ACIA, 2004) stressed that Arctic climate change is a concern of the present and the future. The Adaptation Actions for the Changing Arctic (AACA, 2017) report provides the first comprehensive understanding of the complexity of Arctic change and sustainable development futures. These reports stress the importance of research that addresses the complex intertwining of continued socio-ecological transformation (UNESCO, 2009). Additionally, some Arctic communities are at the forefront of sustainability research and action, as they work to identify Arctic-specific sustainability challenges and define their own sustainability practices and movements due to rapid and ongoing climate (biogeophysical) changes that affect livelihoods and sometimes entire communities.

Numerous studies constitute Arctic sustainability research (ASR). Spanning multiple scales from the local to circumpolar, encompassing the four pillars of sustainability—economic, environmental, social, cultural—and developing theoretical and methodological approaches, this research involves diverse groups of stakeholders, rights holders, and knowledge holders. ASR addresses pressing SES concerns in inter- and transdisciplinary research, contributes to community-based knowledge co-production, and builds connections to policy and practice (Chapin et al., 2009; Petrov et al., 2017). ASR recognizes Indigenous peoples as bearers of sustainability knowledge and the people most affected by environmental change (Gad & Strandsbjerg, 2018). Major contributions to sustainability research by Arctic scholars include, but are not limited to, studies of SES vulnerability (Ford & Pearce, 2010; Melillo, Richmond, & Yohe, 2014; Parlee, 2015), resilience (ARR, 2013, 2016; Armitage, Berkes, Dale, Kocho-Schellenberg, & Patton, 2011; BurnSilver et al., 2017), adaptation to climate and social change (Kofinas et al., 2013; AACA, 2017; Ford & Pearce, 2010), community understandings of sustainability (Crate, 2006; Graybill, 2009; Riedlsperger et al., 2017), engagement of Indigenous knowledge (Kofinas, 2005; Hovelsrud & Smit, 2010; Parlee & Caine, 2018), and cutting-edge methodologies (BurnSilver et al., 2017; Petrov et al., 2016). Arctic sustainability scholars also address transformation of the socio-ecosphere in the Anthropocene, such non-renewable resource extraction (Southcott, Abele, Natcher, & Parlee, 2018; Stammler, 2010, Rodon & Lévesque, 2015; Tysiachniouk & Petrov, 2018); identity, culture, and sustainability (Schweitzer, Fox, Csonka, & Kaplan, 2010; Graybill, 2019); geopolitics (Berkman & Young, 2009; Heininen, 2016); disasters (Bronen & Chapin, 2013); ecosystem stewardship (Chapin et al., 2015); urban development (Orttung, 2016; Taylor et al., 2016); and community capacity building for resilience and adaptation (Ford et al., 2016; Berman et al., 2017; Parlee, 2015).

Methodologies for ASR evolve as sustainability science matures and as the Arctic transforms. We understand methodological transitions to have occurred when the main thrust of research inquiry changes (Figure 1.1). For example, if the impacts of changing environment on humans (so-called "human dimensions" of climate change) was emphasized in the early 2000s (Stammler, 2010), it then transitioned to examining coupled human–natural systems by the decade's end (Chapin et al., 2009) and to social-ecological systems in the next decade (ARR, 2013), thereby indicating interlinked, complex systems related to natural and human vulnerabilities, resiliencies, and adaptive capacities (AACA, 2017). Most recently, state-of-the-art methods are theoretically and experientially informed, community and problem based, and often transdisciplinary in nature, engaging Indigenous and local knowledge to create collaborative, complementary ways of knowing (Kruse et al., 2004; Johnson et al., 2016; Petrov et al., 2016).

How is ASR relevant to other sustainability scholarship? A general critique of Arctic scholarship is its isolation from scholarship about the rest of the world, resulting in Arctic exceptionalism. Resultingly, there are calls to (re)connect Arctic scholarship with global research (Kristoffersen & Langhelle, 2017; Petrov et al., 2017; Anderson et al., 2018). While the Arctic's socio-environmental changes and challenges are unique, knowledge exchange about sustainability beyond the Arctic is desired. Some efforts have already been fruitful (Taylor et al., 2016), and others are forthcoming. What ASR provides are theoretical, methodological, and practical

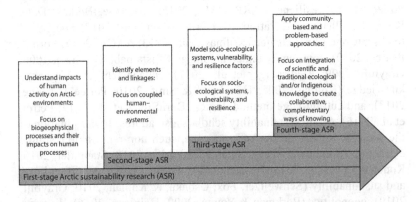

Figure 1.1 Conceptual diagram indicating that ways of addressing sustainability challenges have altered over the past decade and adapted to the specific concerns of the Arctic. Note that there is no end time noted for any of the four stages of Arctic sustainability research; research and practitioner use of each stage continues, adding to the multiplicity of knowledge and applied action toward sustainability in and of the Arctic.

contributions to sustainability science. Examples of the global relevance of ASR include theories of resilience and adaptation, sustainability indicators, development scenarios, community-based and transdisciplinary research methods, and transdisciplinary collaborations with stakeholders, rights holders, and knowledge holders.

Converging knowledge systems

Rapidly changing, complex SESs require Arctic residents to adapt their systems of knowledge to live sustainably, thereby increasing the demand for the convergence of Indigenous and Western knowledge systems. Arctic sustainability research possesses attributes of such convergence science (NRC, 2014) because it targets complex, globally relevant and locally important challenges in highly integrated, inter- and transdisciplinary ways (Petrov et al., 2016). It also may offer effective strategies and solutions for adaptation based on knowledge co-production between Indigenous and local communities and Western scientists.

Indigenous communities are often a major focus of ASR. However, how Indigenous peoples have been included in sustainable development scholarship is often problematic (Gad et al., 2017). While their resilience is emphasized (Vlasova, 2002; ARR, 2016), some have found 'traditional' societies to be the most vulnerable to change (Ford et al., 2015). More recent literature promotes legitimacy of Indigenous knowledge (IK) for sustainability (Hovelsrud & Smit, 2010; Chapin et al., 2009, Lavrillier & Gabyshev, 2017). Various definitions of (IK) traditional knowledge (TK), traditional ecological knowledge (TEK), and local and Indigenous knowledge (LIK) (ICC, n.d.; Arctic Council, 2018; Hirshberg & Petrov, 2014; Brhlíková, 2017) exist, but the foundational understanding is that such knowledge is a self-sustaining knowledge system reflecting generations-long experiences of Indigenous peoples that provides a foundation for individual and collective wellbeing of past, present, and future generations of Arctic Indigenous peoples and holds inherent value, methodologies, functions, and validation processes. IK empowers Arctic communities to advance the understanding, intellectual performance, and management of the Arctic (Behe et al., 2017). Developed over many centuries, IK is a dynamic, adaptive knowledge system that offers unique ways of knowing (Johnson et al., 2016).

This said, the concept of sustainable development is a Western construct. Scholars must maintain constant awareness of sustainability and IK so that sustainability does not become another tool of colonization or romanticization of Indigenous societies (see Petrov et al., 2017; Gad & Strandsbjerg, 2018). Scholarship aiming to decolonize recent sustainability practices engages knowledge co-production *with* Indigenous communities to ensure equity and legitimacy of knowledge systems (Johnson et al., 2016; Alessa et al., 2016; Armitage et al., 2011). While there are gaps between IK, Western science, and decision making, Arctic Indigenous peoples have a substantial knowledge of agriculture, medicine, environment, biodiversity, and other areas of sustainable human existence (Houde, 2007; Huntington, 2011; Uprety, Asselin, Dhakal, & Julien, 2012). Petrov et al. (2017) suggest a major shift toward knowledge co-production approaches that engage IK, emphasizing that understanding sustainability as a normative outcome has caused scholars to see sustainability as a process in which community-informed research may best address complex sustainability challenges. Some Arctic communities understand sustainability as "how communities envision and pursue social and natural wellbeing," in which communities have the right to "drive their own sustainable futures" (Riedlsperger et al., 2017, p. 319). However, improving co-production methodologies and advancing knowledge co-production in an equitable manner remains to be accomplished (Behe et al., 2017).

Toward promoting Arctic sustainability

Arctic sustainability research focuses on SESs, which is a unit of analysis and a way of operationalizing concepts such as resilience, adaptation, robustness, and thrivability (ARR, 2013, 2016). SES is an intertwined system of natural and social phenomena and processes linked by mutual dependencies and dynamic reciprocities. SESs are often understood as being interconnected through ecosystem services that explicitly mediate between social and natural subsystems. As such, there is need to develop an Arctic-focused practice of sustainability. In this book, we adopted a working definition proposed by the Arctic FRontiers Of SusTainability: Resources, Societies, Environments and Development in the Changing North (Arctic-FROST) research coordination network (Petrov, 2014). Developed over a five-year process involving dozens of scholars and Arctic community members, we understand sustainable Arctic development to be *"development that improves the health, well-being and security of Arctic communities and residents while conserving ecosystem structures, functions, and resources."*

This Arctic-specific definition explicitly engages "ecology, socio-environmental justice, and equity, while recognizing the need to live within the supporting ecosystem's limits" (after Agyeman et al., 2003). What's really exciting about this Arctic sustainability definition is the sense that there may be a better way forward for people in this region because of the really hard work that's been done by and for multiple stakeholders in government, in academia, and by local and Indigenous peoples from communities. In defining Arctic sustainability, we maintain the multi-generational nature of this concept. Too often sustainable development is reduced to *ad hoc* activities to resolve immediate problems or to deciding how or when to act instead of actually doing something. Although address-ing emergencies is part of sustainable development, keeping a long-term perspective and working toward sustainability with multiple generations is crucial to success. Here, it's important to remember that sustainable devel-opment is a process and an outcome, and a focus on process rather than on a (normative) outcome may be a more productive, action-oriented approach (Petrov et al., 2017).

Navigating Arctic sustainability in this book

Because the Arctic is a region composed of different types of communi-ties and places and because sustainability is a multifaceted concept, con-ceptualizing Arctic sustainability requires attention to multiple aspects of this region and the people and places in it. Over multiple years working together as a team, the authors and editors of this book have developed a working framework that places ASR on four pillars of sustainability— environment, society, culture, and economy—that include seven critical knowledge domains: sustainable governance, sustainable communities, sustainable environments, sustainable culture, sustainable economies, sus-tainable resource, and sustainable cities (Figure 1.2). Additionally, multiple crosscutting themes inform Arctic sustainability: natural resource devel-opment and management, climate change, biogeophysical transformations, human wellbeing, education and health, gender and socio-environmental justice, Indigenous communities and knowledge, and globalization. Within this framework, trends and drivers of change and, hopefully, of sustainabil-ity for the Arctic manifest themselves within dynamic human–environmen-tal Arctic systems.

This book follows the thematic structuring of seven domains and mul-tiple crosscutting themes (see Figure 1.2). With this framework, we find it important to address Arctic sustainability by formulating the concept of "sustainable [X]" instead of "[X] sustainability." We take this approach to focus on sustainability of a particular domain. For example, we wish to understand sustainable economic systems in the Arctic (i.e., the economy domain of sustainability) rather than economic aspects of sustainability as they transpire across various domains.

Our hope is that this book about Arctic sustainability accomplishes two goals. First, we hope that this book contributes to the growing body of schol-arship about sustainability research and practice for the global community. Second, we hope that our progressive vision for Arctic sustainability may

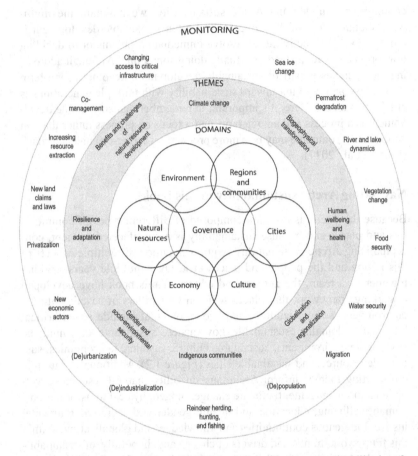

Figure 1.2 Conceptual model used in this book for addressing sustainability in, of, and for the Arctic region. This model provides a visual understanding of how specific issues (shown in the outer ring) relate to the domains (shown in the inner ring) identified in this book are addressed in individual chapters. In this conceptualization, the core components of the sustainability concept are addressed in multiple, overlapping ways so as to ensure that sustainability has been considered from multiple knowledge domains.

help push forward the field of sustainability research and practice in and for the Arctic as well as in other global communities and places.

This book is intended to be the first of two volumes synthesizing the state of knowledge about Arctic sustainability and addresses four realms of sustainability and methods for addressing Arctic sustainability. In Volume I, the domains of economies, resources, cultures, and governance are addressed as well as methodologies for conducting sustainability studies in the Arctic. Volume II will address the domains of communities, cities, environment, and Indigenous perspectives and present a synthesis of monitoring for sustainability in the Arctic. A brief outline of each chapter in Volume I follows.

Chapter 2, "Sustainable Economies in the Arctic" (Joan Nymand Larsen and Lee Huskey): Garnering much attention in global and Arctic-focused literature, a better understanding of the definition(s) and the means of achieving economic sustainability is needed for the Arctic. This chapter develops a usable definition of economic sustainability for the Arctic by synthesizing important elements of the Arctic economy with economic sustainability literature to provide characteristics of sustainability and hypotheses about successful development. An operational definition reflecting scale, perspective, and the Arctic-specific region allows for examination of sustainability successes and failures, suggesting possible goals and policy options for achieving sustainable economic development.

Chapter 3, "Culture and Sustainability" (Susanna Gartler, Vera Kuklina, and Peter Schweitzer): Despite change having long characterized Arctic cultures, the rapidity and scale of current changes (climatic, social) endanger Arctic cultures as we know them. Global environmental change threatens traditional subsistence activities and the physical existence of some settlements, but forces alluded to under the label of globalization bring into question the continuation of local cultural practices (Schweitzer et al., 2010). While monitoring and analysis of Arctic cultural processes is performed by multiple stakeholders (scholars, activists, practitioners), this chapter synthesizes knowledge about cultural sustainability with the goals of understanding present and potential trajectories of cultural development and identifying ways to foster cultural vitality.

Chapter 4, "Sustainable Resources" (Chris Southcott): Rapid land use change from industrial development and climate change challenge sustainability of Arctic SESs and natural resources. Controversial decisions about sustaining Arctic natural resources reflect underlying societal dilemmas inherent in sustainable development choices. Interaction of climate change and land use change challenges the sustainability of resources, highlighting the need to focus on sustainable resource use and maintaining ecosystem services. Drawing on diverse resource stewardship cases to synthesize common and unique patterns of SES interactions, this chapter explores if and how sustainability can be achieved in the changing Arctic.

Chapter 5, "Governance for Arctic Sustainability" (Gary N. Wilson, Gail Fondahl, and Klaus Georg Hansen): The Arctic is home to a complex and interconnected web of local, regional, national, and global governance institutions. The Arctic is also at the forefront of development in Indigenous governance, with regions such as Greenland on the cusp of statehood and Indigenous peoples such as the Sámi and the Inuit achieving considerable autonomy from their host states. In addition to the traditional institutions of government, powerful non-governmental organizations play an important role in Arctic governance. This chapter examines best practices in Arctic governance and develops case studies that illustrate successful and unsuccessful elements of governance in the Arctic. Particular attention is paid to Indigenous forms of governance because Indigenous communities are where the most pressing challenges lie.

Chapter 6, "Methodological Challenges and Innovations in Arctic Community Sustainability Research" (Gary Kofinas, Shauna BurnSilver, and Andrey N. Petrov): This chapter reviews methodological challenges and advancements of ASR, focusing on community sustainability. The authors describe methodological shifts and trends, such as transitions to inter- and transdisciplinarity and knowledge co-production. Innovative strategies for transcending problems in ways that contribute to collaboration, knowledge production, and social learning are identified by examining five key challenges and arenas for research innovation and action: making research relevant to communities, co-producing knowledge, accounting for scale and linkage, developing indicators, and realizing praxis.

References cited

ACIA. (2004). *Impacts of a warming Arctic: Arctic climate impact assessment. ACIA overview report.* Retrieved from www.amap.no/documents/doc/arctic-arctic-climate-impact-assessment/796

Agyeman, J., Bullard, R., & Evans, B. (Eds.). (2003). *Just sustainabilities: Development in an unequal world.* Cambridge, MA: MIT Press.

Alessa, L., Kliskey, A., Gamble, J., Fidel, M., Beaujean, G., & Gosz, J. (2016). The role of Indigenous science and local knowledge in integrated observing systems: Moving toward adaptive capacity indices and early warning systems. *Sustainability Science, 11*(1), 91–102.

AMAP. (2017). *Adaptation actions for a changing Arctic: Perspectives from the Barents region.* Oslo, Norway: Arctic Monitoring and Assessment Programme (AMAP).

Anderson, S., Strawhacker, C., Presnall, A., Butler, V., Etnier, M., Petrov, A., & Yamin-Pasternak, S. (2018). *Arctic horizons: Final report.* Washington, DC: Jefferson Institute.

Arctic Council. (2013). *Arctic resilience interim report 2013.* Retrieved from www.arctic-council/arr

Arctic Council. (2016). *Arctic resilience report* (M. Carson & G. Peterson, Eds.). Retrieved from www.arctic-council.org.arr

Arctic Council. (2018). *Ottawa traditional knowledge principles.* Retrieved from www.arcticpeoples.com/our-work-2#traditional-knowledge-1

Armitage, D., Berkes, F., Dale, A., Kocho-Schellenberg, E., & Patton, E. (2011). Co-management and the co-production of knowledge: Learning to adapt in Canada's Arctic. *Global Environmental Change, 21*(3), 995–1004.

Armitage, D., Berkes, F., & Doubleday, N. (Eds.). (2010). *Adaptive co-management: Collaboration, learning, and multi-level governance.* Vancouver: UBC Press.

Åtland, K. (2011). Russia's Armed Forces and the Arctic: All Quiet on the Northern Front?, *Contemporary Security Policy, 32*(2), 267–285.

Bawaka Country, Suchet-Pearson, S., Wright, S. W., Lloyed, K., & Burrawanga, L. (2013). Caring as country: Towards an ontology of co-becoming in natural resource management. *Asia Pacific Viewpoint, 54*(2), 185–197.

Behe, C., Blind, A., Johnson, N., Mack, L., Mathisen, S., & Petrov, A. (2017). Statement of the ICASS IX Indigenous knowledge roundtable. *Northern Notes, 48*, 10

Berkes, F., Colding, J., & Folke, C. (2003). *Navigating socio-ecological systems: Building resilience for complexity and change.* Cambridge: Cambridge University Press.

Berkman, P. A., & Young, O. R. (2009). Governance and environmental change in the Arctic Ocean. *Science, 324*(5925), 339–340.

Berman, M., Kofinas, G., BurnSilver, S., Fondahl, G., & Wilson, G. N. (2017). Measuring community adaptive and transformative capacity in the Arctic context. In *Northern sustainabilities: Understanding and addressing change in the circumpolar world* (pp. 59–75). Cham: Springer.

Braungart, M., & McDonough, W. (2002). *Cradle to cradle: Remaking the way we make things.* New York: North Point Press.

Brhlíková, L. (2017). *Traditional and local knowledge in the Arctic.* Retrieved from https://projekter.aau.dk/projekter/files/262031076/Traditional_and_Local_Knowledge_in_the_Arctic.pdf

Bronen, R., & Chapin, F. S. (2013). Adaptive governance and institutional strategies for climate-induced community relocations in Alaska. *Proceedings of the National Academy of Sciences, 110*, 9320–9325.

Brundtland, G. H. (1987). *Our common future: Report of the world commission on environment and development.* Oxford: Oxford University Press.

BurnSilver, S., Boone, R., Kofinas, G., & Brinkman, T. (2017). Modeling tradeoffs in a rural Alaska economy: Hunting, working and sharing in the face of economic and ecological change. In M. Hegmon (Ed.), *The give and take of sustainability: Archaeological and anthropological perspectives on tradeoffs.* Cambridge: Cambridge University Press.

Cameron, E. (2012). Securing indigenous politics: A critique of the vulnerability and adaptation approach to the human dimensions of climate change in the Canadian Arctic. *Global Environmental Change, 22*(1), 103–114.

Chapin, F. S., Kofinas, G. P., & Folke, C. (2009). *Principles of ecosystem stewardship: Resilience-Based natural resource management in a changing world.* New York, NY: Springer Science.

Chapin, F. S. III, Sommerkorn, M., Robards, M. D., & Hillmer-Pegram, K. (2015). Ecosystem stewardship: A resilience framework for arctic conservation. *Global Environmental Change, 34*, 207–217.

Clark, W. C., & Dickson, N. M. (2003). Sustainability science: The emerging research program. *National Academy of Sciences, 100*(14), 8059–8061.

Crate, S. A. (2006). *Cows, kin, and globalization: An ethnography of sustainability.* Altamira: Rowman.

Denzin, N., Norman, K., Lincoln, Y. S., & Tuhiwai, L. (2008). *Handbook of critical and indigenous methodologies.* Los Angeles: Sage.

DeVries, B. J. M. (2013). *Sustainability science.* Cambridge: Cambridge University Press.

Dinero, S. C. (2011). Indigenous perspectives of climate change and its effects upon subsistence activities in the Arctic: The case of the Nets'aii Gwich'in. *Geojournal, 78*(1), 117–137.

Emmerson, C. (2010). *The future history of the Arctic.* New York, NY: Public Affairs.

Folke, C., Hahn, T., Olsson, P., & Norberg, J. (2005). Adaptive governance of social-ecological systems. *Annual Review of Environment and Resources, 30,* 441–473.

Fondahl, G., & Wilson, G. N. (Eds.). (2017). *Northern sustainabilities: Understanding and addressing change in the circumpolar world.* New York, NY: Springer.

Forbes, B. (2013). Cultural resilience of social-ecological systems in the Nenets and Yamal-Nenets Autonomous Okrugs, Russia: A focus on reindeer nomads of the tundra. *Ecology and Society, 18*(4), 36.

Forbes, B. C., & Stammler, F. (2009). Arctic climate change discourse: The contrasting politics of research agendas in the West and Russia. *Polar Research, 28*(1), 28–42.

Ford, J. D., McDowell, G., & Pearce, T. (2015). The adaptation challenge in the Arctic. *Nature Climate Change, 5*(12), 1046.

Ford, J. D., & Pearce, T. (2010). What we know, do not know, and need to know about climate change vulnerability in the western Canadian Arctic: A systematic literature review. *Environmental Research Letters, 5*(1), 014008.

Ford, J. D., Stephenson, E., Cunsolo Willox, A., Edge, V., Farahbakhsh, K., Furgal, C., & Austin, S. (2016). Community-based adaptation research in the Canadian Arctic. *Wiley Interdisciplinary Reviews: Climate Change, 7*(2), 175–191.

Funtowicz, S. O., & Ravetz, J. R. (1991). A new scientific methodology for global environmental issues. *Ecological Economics: The Science and Management of Sustainability, 10,* 137.

Gad, U. P., Jakobsen, U., & Strandsbjerg, J. (2017). Politics of sustainability in the Arctic: A research agenda. In G. Fondahl & G. N. Wilson (Eds.), *Northern sustainabilities: Understanding and addressing change in the circumpolar world* (pp. 13–23). Cham: Springer.

Gad, U. P., & Strandsbjerg, J. (Eds.). (2018). *The politics of sustainability in the Arctic: Reconfiguring identity, space, and time.* New York, NY: Routledge.

Graybill, J. K. (2007). Continuity and change: (re)constructing environmental geographies in late Soviet and post-Soviet Russia. *Area, 39*(1), 6–19.

Graybill, J. K. (2009). Places and identities on Sakhalin Island: Situating the emerging movements for 'Sustainable Sakhalin.' In J. Agyeman & E. Ogneva-Himmelberger (Eds.), *Environmental justice of the former Soviet Union.* Boston: MIT Press.

Graybill, J. K. (2013). Imagining resilience: Situating perceptions and emotions about climate change on Kamchatka, Russia. *GeoJournal, 78,* 817–832.

Graybill, J. K. (2019). Emotional environments of energy extraction in Russia. *Annals of the American Association of Geographers, 109*(2), 382–394.

Haalboom, B., & Natcher, D. C. (2012). The power and peril of "vulnerability": Approaching community labels with caution in climate change research. *Arctic, 65*(3), 319–327.

Heininen, L. (Ed.). (2016). *Future security of the global Arctic: State policy, economic security and climate.* London: Springer.

Heininen, L., & Southcott, C. (Eds.). (2010). *Globalization and the circumpolar north.* Fairbanks: University of Alaska Press.

Heininen, L. E.-P., & Plouffe, H. J. (2016). *Arctic yearbook: The Arctic Council: 20 Years of regional cooperation and policy-shaping.* Reykjavik: NRF.

Hirshberg, D., & Petrov, A. (2014). Education and human capital. In J. N. Larsen & G. Fondahl (Eds.), *Arctic human development report: Regional processes and global linkages* (pp. 349–399). Copenhagen: TemaNord.

Holling, C. S., & Gunderson, L. H. (Eds.). (2002). *Panarchy: Understanding transformations in human and natural systems.* Washington, DC: Island Press.

Houde, N. (2007). The six faces of traditional ecological knowledge: Challenges and opportunities for Canadian co-management arrangements. *Ecology and Society, 12*(2), 34.

Hovelsrud, G. K., Poppel, B., Van Oort, B., & Reist, J. D. (2011). Arctic societies, cultures, and peoples in a changing cryosphere. *AMBIO, 40*(1), 100–110.

Hovelsrud, G. K., & Smit, B. (Eds.). (2010). *Community adaptation and vulnerability in Arctic regions.* Dordrecht: Springer.

Huntington, H. P. (2011). Arctic science: The local perspective. *Nature, 478*(7368), 182.

Indigenous Circumpolar Council. (n.d.). *Indigenous knowledge.* Retrieved from www.inuitcircumpolar.com/icc-activities/environment-sustainable-development/indigenous-knowledge/

Jacob, K., Blake, R., Horton, R., Bader, D. A., & O'Grady, M. (2010). Indicators and monitoring. *Annals of the New York Academy of Sciences,* 1196, 127–142.

Johnson, J. T., Howitt, R., Cajete, G., Berkes, F., Louis, R. P., & Kliskey, A. (2016). Weaving indigenous and sustainability sciences to diversify our methods. *Sustainability Science, 11*(1), 1–11.

Johnson, N., Behe, C., Danielsen, F., Krümmel, E. M., Nickels, S., & Pulsifer, P. L. (2016). *Community-based monitoring and indigenous knowledge in a changing Arctic.* Ottawa: Inuit Circumpolar Council.

Kates, R. W. (2011). What kind of a science is sustainability science? *Proceedings of the National Academy of Sciences, 108,* 19449–19450.

Keskitalo, C. (2004). *Negotiating the Arctic: The construction of an international region.* London: Routledge.

Keskitalo, E. C. H., Dannevig, H., Hovelsrud, G. K., West, J. J., & Swartling, A. G. (2011). Adaptive capacity determinants in developed states: Examples from the Nordic countries and Russia. *Regional Environmental Change, 11,* 579–592.

Kofinas, G. P. (2005). Caribou hunters and researchers at the co-management interface: Emergent dilemmas and the dynamics of legitimacy in power sharing. *Anthropologica,* 179–196.

Kofinas, G. P., Clark, D., Hovelsrud, G. K., Alessa, L., Amundsen, H., Berman, M., ... Olsen, J. (2013). Adaptive and transformative capacity. In *Arctic resilience interim report 2013*. Stockholm, Sweden: Arctic Council.

Kristoffersen, B., & Langhelle, O. (2017). Sustainable development as a global-arctic matter: Imaginaries and controversies. In K. Keil & S. Knecht (Eds.), *Governing Arctic Change*, 21–41. London: Palgrave Macmillan.

Krupnik, I., Allison, I., Bell, R., Cutler, P., & Hik, D. (2011). Understanding earth's polar challenges: International polar year 2007–2008. In *Governing Arctic change* (pp. 21–41). London: Palgrave Macmillan.

Kruse, J. A., White, R. G., Epstein, H. E., Archie, B., Berman, M., Braund, S. R., & Flanders, N. (2004). Modeling sustainability of arctic communities: An interdisciplinary collaboration of researchers and local knowledge holders. *Ecosystems*, 7(8), 815–828

Laidler, G. J., Ford, J. D., Gough, W. A., et al. (2009). *Climatic Change*, 94: 363. https://doi.org/10.1007/s10584-008-9512-z

Laurie, T. (2012). Epistemology as politics and the double-bind of border thinking: Levi-Strauss, Deleuze, Mignolo. *Journal of Multidisciplinary International Studies*, 9(2), 1–20.

Lavrillier, A., & Gabyshev, S. (2017). *An Arctic indigenous knowledge system of landscape, climate, and human interactions*. Fürstenberg: Kulturstiftung Sibirien/SEC Publications.

Lindstad, H., Bright, R. M., & Strømman, A. H. (2016). Economic savings linked to future Arctic shipping trade are at odds with climate change mitigation. *Transport Policy*, 45, 24–30.

Loranty, M. M., & Goetz, S. J. (2012). Shrub expansion and climate feedbacks in Arctic tundra. *Environmental Research Letters*, 7(1).

Lyons, O. (1980). An Iroquois perspective. In *American Indian environments: Ecological issues in native American history* (pp. 171–174). Syracuse: Syracuse University Press.

Ma, J., Hung, H., Tian, C., & Kallenborn, R. (2011). Revolatilization of persistent organic pollutants in the Arctic induced by climate change. *Nature Climate Change*, 1(5), 255–260.

Macias-Fauria, M., Forbes, B.C., Kumpula, T., & Zetterberg, P. (2012). Eurasian Arctic greening reveals teleconnections and the potential for novel ecosystems. *Nature Climate Change*, 2, 613–618.

Meadows, D. H. (2008). *Thinking in systems: A primer*. Vermont: Chelsea Green.

Melillo, J. M., Richmond, T. C., & Yohe, G. W. (2014). *Climate change impacts in the United States: The third national climate assessment*. Retrieved from https://nca2014.globalchange.gov/downloads

Mignolo, W. (2011). *The darker side of western modernity: Global futures, decolonial options*. Durham: Duke University Press.

Moiseev, N. N. (1990). *Man and noosphere*. Moscow: Young Guard.

National Research Council. (2014). *The Arctic in the Anthropocene: Emerging research questions*. Washington, DC: National Academies Press.

Oestreicher, J. S., Buse, C., Brisbois, B., Patrick, R., Jenkins, A., Kingsley, J., & Fatorelli, L. (2018). Where ecosystems, people and health meet: Academic traditions and emerging fields for research and practice. *Sustainability in Debate*, *9*(1), 45–65.

Orttung, R. W. (Ed.). (2016). *Sustaining Russia's Arctic cities: Resource politics, migration, and climate change*. New York, NY: Berghahn Books

Parlee, B. L. (2015). Avoiding the resource curse: Indigenous communities and Canada's oil sands. *World Development, 74*, 425–436.

Parlee, B. L., & Caine, K. J. (2018). *When the caribou do not come: Indigenous knowledge and adaptive capacity*. Vancouver: UBC Press.

Parr, A. (2012). *Hijacking sustainability*. Cambridge: MIT Press.

Petrov, A. N. (2014). Sustainability and sustainable development in the Arctic: Meanings and means. *Proceedings of the Arctic-FROST workshop September 18–19*. Anchorage, Alaska.

Petrov, A. N., BurnSilver, S., Chapin, F. S. III, Fondahl, G., Graybill, J. K., Keil, K., . . Schweitzer, P. (2016). Arctic sustainability research: Toward a new agenda. *Polar Geography, 39*(3), 165–178.

Petrov, A. N., BurnSilver, S., Chapin, F. S. III, Fondahl, G., Graybill, J. K., Keil, K., . . Schweitzer, P. (2017). *Arctic sustainability research: Past, present and future*. London: Routledge.

Pitt, D., & Samson, P. R. (1999). *The biosphere and noosphere reader: Global environment*. New York, NY: Routledge.

Quijano, A. (2007). Coloniality and modernity/rationality. *Cultural Studies, 21*(2–3), 168–178.

Raynolds, M. K., Walker, D. A., Munger, C. A., Vonlanthen, C. M., & Kade, A. N. (2008). A map analysis of patterned-ground along a North American Arctic transect. *Journal of Geophysical Research*, 113, G03S03. https://doi.org/10.1029/2007JG000512

Riedlsperger, R., Goldhar, C., Sheldon, T., Bell, T., Fondahl, G., & Wilson, G. N. (2017). Meaning and means of "sustainability": An example from the Inuit Settlement Region of Nunatsiavut, Northern Labrador. In *Northern sustainabilities: Understanding and addressing change in the circumpolar world* (pp. 317–336). Cham: Springer.

Rodon, T., & Lévesque, F. (2015). Understanding the social and economic impacts of mining development in Inuit communities: Experiences with past and present mines in Inuit Nunangat. *Northern Review, 41*, 13–39.

Schweitzer, P., Fox, S. I., Csonka, Y., & Kaplan, L. (2010). Cultural wellbeing and cultural vitality. In *Arctic social indicators*. Copenhagen: Nordic Council of Ministers.

Shaw, D. J. B., & Oldfield, J. D. (2006). V.I. Vernadsky and the noosphere concept: Russian understandings of society—nature interaction. *Geoforum, 37*, 145–154.

Smit, B., Wandel, J. (2006) Adaptation, adaptive capacity and vulnerability. *Global Environmental Change*, 16, 282–292.

Smith, L. (2011). *The world in 2050: Four forces shaping civilization's northern future*. New York, NY: Plume.

Southcott, C., Abele, F., Natcher, D., & Parlee, B. (Eds.). (2018). *Resources and sustainable development in the Arctic*. New York, NY: Routledge.

Stammler, F. (2010). Our movement to retire the term "human dimension" from the Arctic science vocabulary. *Northern Notes, 34*, 7–13.

Swyngedouw, E. (2010). The trouble with Nature: Ecology as the New Opium for the Masses. In P. Healey & J. Hillier (Eds), *Conceptual Challenges for Planning Theory*. Aldershot: Ashgate.

Taylor, A., Carson, D. B., Ensign, P. C., Huskey, L., Rasmussen, R. O., & Saxinger, G. (Eds.). (2016). *Settlements at the edge: Remote human settlements in developed nations*. Cheltenham, UK: Edward Elgar Publishing.

Tennberg, M. (1998). *The Arctic Council: A study in governmentality*. Rovaniemi: University of Lapland.

Turner, B. L., Kasperson, R. E., Matson, P. A., McCarthy, J. J., Corell, R.W., Christensen, L., Eckley, N., Kasperson, J. X., Luers, A., Martello, M. L., Polsky, C., Pulsipher, A. & Schiller, A. A framework for vulnerability analysis in sustainability science. *Proceedings of the National Academy of Sciences*, *100*, 14, 804–807.

Tysiachniouk, M. S., & Petrov, A. N. (2018). Benefit sharing in the Arctic energy sector: Perspectives on corporate policies and practices in Northern Russia and Alaska. *Energy Research & Social Science*, *39*, 29–34.

UNEP. (1991). *Caring for the earth: A strategy for sustainable living*. Gland, Switzerland: UNEP.

UNESCO. (2009). *Climate change and Arctic sustainable development: Scientific, social, cultural, and educational challenges*. Paris: UNESCO.

Uprety, Y., Asselin, H., Dhakal, A., & Julien, N. (2012). Traditional use of medicinal plants in the boreal forest of Canada: Review and perspectives. *Journal of Ethnobiology and Ethnomedicine*, *8*(1), 7.

Vernadsky, V. I. (2002). *Biosfera i noösfera*. Moscow: Airis Press.

Vlasova, T. K. (2002). Human impacts on the tundra-taiga zone dynamics: The case of the Russian lesotundra. *AMBIO*, 30–36.

Vlasova, T. K., & Volkov, S. (2016). Towards transdisciplinarity in Arctic sustainability knowledge co-production: Socially-oriented observations as a participatory integrated activity. *Polar Science*, *10*(3), 425–432.

Voinov, A., Bromley, L., Kirk, E., Korchak, A., Farley, J., Moiseenko, T., Krasovskaya, T., Makarova, Z., Megorsm, V., Selin, V., Kharitonova, G., & Edson, R. (2004). Understanding human and ecosystem dynamics in the Kola Arctic: A participatory integrated study. *Arctic*, *57*(4), 375–388.

Walker, T. R., Habeck, J. O., Karjalainen, T. P., Virtanen, T., Solovieva, N., Jones, V., Kuhry, P., Ponomarev, V. I., Mikkola, K., Nikula, A., Patova, E., Crittenden, P. D., Young, S. D., & Ingold, T. (2006). Perceived and measured levels of environmental pollution: Interdisciplinary research in the subarctic lowlands of northeast European Russia. *AMBIO – A Journal of the Human Environment*, *35*(5), 220–228.

Walker, S., Salt, D. (2006). Resilience thinking: Sustaining ecosystems and people in a changing world. Washington DC: Island Press.

Wilson, E. (2003). Freedom and loss in a human landscape: Multinational oil exploitation and the survival of reindeer herding in north-eastern Sakhalin, the Russian Far East. *Sibirica: The Journal of Russia in Asia and the North Pacific*, *3*(1), 21–48.

Young, O. R. (2010). Arctic governance—pathways to the future. *Arctic Review on Law and Politics*, *1*(2), 164–185.

2 Sustainable economies in the Arctic

Joan Nymand Larsen and Lee Huskey

Introduction

The material wellbeing of Arctic residents is determined by their consumption and control over goods and services (Larsen & Huskey, 2010). A region's economy is the source of a community's material wellbeing, so a sustainable economy is necessary to long-term wellbeing. Unfortunately, the nature of the Arctic economy makes sustainability difficult to achieve. A sustainable economy provides for the material wellbeing of current residents while not compromising the wellbeing of future residents (Brundtland, 1987). A sustainable economy must be organized to maintain its current economic base and to replace that base when faced with external or internal challenges. This means those making economic decisions must reduce the current resource use, invest in alternatives, and save current income. These are difficult decisions that require balancing current and future wellbeing (Pezzey & Toman, 2002a, 2002b).

There are two primary drivers of the Arctic economy (Huskey, Maenpaa, & Pelyasov, 2014). One is the harvest and production of northern natural resources for local consumption and sale in external markets. Traditional or subsistence harvesting provides real income for residents, especially those living in smaller, more remote communities (Holen et al., 2015). Timber, diamonds, gold, fish, natural gas, and oil are produced in the North but mostly sold elsewhere. While both types of production are affected by local resource availability, production for the international economy is influenced by businesses and government decisions makers in other regions.

The economic effect of resource production on the Arctic economy reflects different production methods (Huskey, 2010; Huskey & Southcott, 2016). Production for external markets is supported by large, concentrated, capital-intensive facilities, whereas production for local consumption often occurs in small, scattered groups using traditional methods. The scale, capital needs, and skills to produce for export require external resources (Duhaime, 2004; Larsen & Huskey, 2015; Huskey et al., 2014).

The second driver is public spending, which may be more important to economic wellbeing than natural resource production (Duhaime et al., 2017). Spending by local, regional, and national governments creates government, construction, and service industry jobs. Governments provide income through transfer payments, such as public assistance or age-related support programs. The public sector contributes to residents' real income by providing services below their real costs; subsidized health care or housing adds to economic wellbeing (Glomsrød, Mäenpää, Lindholt, McDonald, & Goldsmith, 2017; Larsen & Huskey, 2010; Larsen, 2004,b). Public spending often depends on national or regional government decisions.

An economy can be described as a region bound together by its network of economic relationships (Fox & Kumar, 1965). Community and regional economies may have similar economic structure but not represent an integrated economy. Concerns about an economy's sustainability are questions about local economies. The local focus of the sustainability question means that scale matters. Community sizes across the North range from villages of a few hundred to cities of half a million (Heleniak, 2014; Aarsæther, Riabova, & Bærenholdt, 2004). The varieties of scale, history, and environment must be considered to understand Arctic economic sustainability.

Challenges to the sustainability of Arctic economies

A single prescription for the sustainability of Arctic economies doesn't exist, but they share many features that challenge sustainability (Huskey, 2011; Bone, 2009). We describe five characteristics that challenge the sustainability of Arctic economies and limit their ability to maintain current or prepare for future material wellbeing. We show how some Arctic economies respond to these challenges, offering hope for a more sustainable future.

Remoteness

Arctic economies are remote in multiple ways. Arctic populations and resources are far from external markets, sources of supply, and centers of decision making, which increases production cost. The costs of transportation, inventory, labor, and infrastructure increase with distance from more developed regions (Huskey, 2010; Bone, 2009). Because of sparse settlement, northern economies are also remote from each other, which limits linkages between local economies. Additionally, resources valuable to international markets are spread thinly over the North (Tussing, 1984), meaning existing communities, perhaps established near subsistence resources, may be far from resource production sites.

Remoteness limits sustainable economies in two significant ways. First, economic exports from the North must meet the market test (they must earn more than they cost to produce and deliver to markets). The high costs of access limit economic opportunities available to Arctic economies. Second, the lack of linkages among communities limits how economies can adjust to decline. Commuting and migration are responses available to less remote communities but are difficult for residents of remote places.

While remoteness characterizes the Arctic, it is not a constant. Where a history of economic development and infrastructure investment exists, such as in the European north where road networks connect places, Arctic economies are less remote. Connections with more developed regions reflect historic economic activity or explicit government policies (Rasmussen, 2011). The scale and abundance of particular resource deposits may also overcome the limits of remoteness. For example, Alaska's North Slope petroleum deposits were rich enough to support the construction of an 800-mile pipeline.

Narrow economic base

The local resource base supports both traditional production and production for external markets. The traditional economy usually depends on the region's resources for its livelihood. In contrast, resource production for external markets is often based on fewer resources. For example, oil and gas production dominate Alaskan and Russian Arctic economies; mining dominates in the Canadian North (Huskey et al., 2014).

A narrow economic base challenges creation of a sustainable economy because few alternatives exist when a dominant industry collapses. Concentrated production of resources for international markets increases instability and fluctuations in local economies (Larsen, 2004). High costs make Arctic production the first to close when prices fall (Sugden, 1982), and local instability ensues when jobs, population, business activity, and government revenue decrease (Larsen & Huskey, 2015).

Not all Arctic economies are vulnerable to fluctuations in single-resource production. Conditions may support the growth of more than one resource industry, such as fishing and timber production in Alaska or farming and timber in the European North. Some Arctic regions have developed industries independent of resource production, such as the technology-focused industry in northern Finland and Sweden, finance in Iceland, and pan-Arctic tourism. Some Arctic communities, such as Nome, Alaska, or Umeå, Sweden, developed as centers of services, transportation, and administration

(Huskey & Taylor, 2016). Finally, Petrov (2016, 2007) notes creative or knowledge-based economic growth. These industries stabilize northern economies because their activities and market cycles differ from natural resource production (Larsen & Huskey, 2015).

External decision makers

Historically, decisions about using northern resources are made outside the region. National and regional governments determine which local resources are produced and how. External government decisions affect traditional and external market resource production. Decisions about Arctic resource use are also made in boardrooms of the world's business capitals. The scale and technology used in Arctic resource extraction projects require investment and expertise of international resource firms, and investment decisions reflect international market conditions and the availability of investment opportunities in other regions.

The public sector in the Arctic depends on external decision makers because of funds needed from higher levels of government. National economic conditions, laws, and regulations influence national government spending and transfers to local economies. National objectives and budgets influenced the economic growth of the Russian North for most of the 20th century; depopulation of this region in the 1990s reflected changes in national policies (Hill & Gaddy, 2003; Heleniak, 1999).

It is difficult to maintain a sustainable local economy when the decisions affecting it are made externally. The economic wellbeing of or the costs imposed on local populations are seldom considered, and the consequences for local and regional economies played a small part in the fluctuating decisions about petroleum development in the Arctic National Wildlife Refuge (Fountain & Friedman, 2017) or the European Union's ban on the imports of seal products (Sellheim, 2015; Myers, 2000). External decisions may not consider potential conflicts between mining or oil production and local traditional uses or fishing and herding industries (Eira et al., 2008; Hermann et al., 2014; Parlee, 2015). Additionally, external control and investment means that most revenue flows out of the North as taxes, royalties, incomes, and rents—leaving only small contributions for local or regional economic wellbeing.

Recent institutional change across the Arctic has increased the role of local residents in resource production decisions and promoted sustainability. Establishing state and regional governments in Alaska, northern Canada, and Greenland provides local residents increased control of resource production and government spending (Eira et al., 2008; Fidler, 2009; Hansen, Vanclay, Croal, & Skjervedal, 2016). Land claims settlements and impact benefit agreements in North America gives residents ownership-like claims to resources, allowing them to benefit from and control development (Larsen & Huskey, 2015). Local and regional institutional changes allow regions such as the North Slope Borough in Alaska to save for future economic downturns (Huskey, 2017).

Poverty

Poverty and limited economic opportunity exist across the Arctic (Duhaime et al., 2017; Huskey, et al., 2014). Addressing the economic needs of current residents may mean pursuing resource or government projects that depend on external support or have limited economic life, which makes achieving long-term sustainability difficult. Resource projects and government spending may increase community populations above levels that the economy can support when a project ends (Knapp & Huskey, 1988). Arctic economies may face a "people versus place prosperity" development problem (Bolton, 1992): they must choose between helping today's population versus creating a long-term sustainable economy.

Other responses to limited economic opportunity may be more sustainable. The connection between resource projects and local economic well-being is stronger in regions where institutional change gives residents more control over resource development. Across the Arctic, residents have responded to limited economic opportunity by moving (Rasmussen, 2011; Heleniak, 2014). When residents are prepared for the economy in other places, migration may be an individual solution to sustainability. However, migration as a solution is limited if residents are unprepared for the outside job markets or if the social, cultural, and monetary costs of moving are prohibitive.

Climate change

Predicting the effect of climate change on Arctic economic sustainability is difficult. Climate change may reduce the limits imposed by remoteness (Larsen & Huskey, 2015). Warming may allow ice-free travel in Arctic seas and lower the cost of access to and development of some Arctic resources (Prowse et al., 2009). However, bad weather and extended and uncertain schedules may increase development costs and uncertainties.

Any benefits of climate change are uncertain and may not outweigh the costs (Hovelsrud, Poppel, Van Oort, & Reist, 2011; Larsen et al., 2014). While climate change may open Arctic seas for transportation and continental shelf development, it may make it costlier to develop terrestrial resources. Changes in flooding, permafrost, and snow cover will increase production costs even in areas with significant current resource activity. Thawing permafrost may destabilize infrastructure, such as roads, pipe-lines, runways (Prowse et al., 2009; Larsen et al., 2014), and settlements.

Climate change also threatens the traditional economy (Ford et al., 2008, 2006; Larsen et al., 2014; Hovelsrud et al., 2011) by changing productivity and locations of fish and wildlife resources, increasing pollution from Arc-tic shipping, and other ecological challenges. As movement across land and sea becomes more difficult and dangerous, the cost of harvesting traditional resources is likely to rise.

Cases

Case studies illustrate the challenges of economic sustainability in the Arc-tic. We provide examples from the literature of long-term sustainability, the resilience of specific places amid change, and examples of unsustainable economies. Different scales, histories, and environments are considered to understand the limits and responses to possible economic sustainability. We illustrate the five characteristics that challenge sustainability and indicate how places may provide for residents' material wellbeing.

Greenland

Since becoming a self-governing part of the Danish Kingdom in 2009, Greenland seeks greater economic independence from Denmark (Hansen et al., 2016; Nielsen, 2013). Large-scale resource development, especially hydrocarbon exploration and uranium and iron ore mining, may offer greater economic independence and a more self-reliant economy (Pop-pel, 2018; Bjørst, 2017; Wilson, 2016; Nielsen, 2013; Hansen et al., 2016; Trump, Kadenic, & Linkov, 2018; Andersen, 2015; Rasmussen, Roto, & Hamilton, 2014). Recent policymaking promotes mining and the Greenlan-dic government is attempting to attract mining companies (Nuttall, 2013). In 2013, UK-based London Mining was granted a development and exploi-tation license for the Isua iron ore mining project. Recently, the parliament repealed Greenland's zero-tolerance policy on uranium mining because it could potentially alleviate Greenland's dependence on Denmark by build-ing economic self-sufficiency (Bjørst, 2017, p. 31).

While mining development represents a potential income source, ques-tions arise about ensuring that economic gains accrue to Greenlanders. Mining activities, oil exploration, and large-scale industrial development plans have provoked political and social debates about this development for society and environment, the absence of appropriate public participation and consultation, decision-making and regulatory processes, the impacts of extractive industries on hunting and fishing activities, the shortcomings of social and environmental impact assessments, and the possible influx of thousands of foreigners for construction and operational phases of megaprojects (Nuttall, 2013, 2012). Moreover, exploiting mineral resources could pressure public finances, including training, schooling, and infrastructure development costs (Nielsen, 2013).

Resource extraction is never risk free. Hansen and Kørnøv (2010) present a value-rational view of the impact assessment of megaindustry in Greenland's planning and policy context. They note challenges to Greenland's impact assessment (IA) system and discuss how it could contribute to securing environmental management and sustainable development. Similarly, Hansen et al. (2016) argue for social IAs by or on behalf of government and prior to project planning. Regarding the Isua project, Trump et al. (2018) argue for decision analysis that includes environmental, economic, and social dimensions to identify policies and project proposals. Wilson (2016) highlights the need to develop a broader economy that includes nonresource extractive industries and implementing legislation and governance structures to handle the emerging resource economy with clear principles, commitments, and guidance on public consultations (Wilson, 2016, p. 75).

Alaska's North Slope and Prudhoe Bay

Oil development, begun at Prudhoe Bay in 1968, has allowed local residents to capture significant economic benefits that minimize the economic leakage of rents and income from the region, made possible by institutional arrangements with the North Slope Borough government and the Arctic Slope Regional Corporation (Huskey, 2017, p. 2). The North Slope case shows that institutional arrangements are critical to successful outcomes and that Arctic economies can avoid the resource curse and instead capture benefits for residents despite single-industry development. Lessons from the North Slope include the importance of institutional arrangements; balanced growth with non-resource development; the protection of historical advantage, including protection of subsistence livelihoods; and remembering to save. The North Slope established a permanent fund for future use of tax revenues (Huskey, 2017, p. 9).

North Slope oil development, while successful (Kruse, 2010), illustrates persistent challenges related to single-resource dependency. Knapp (2015) argues that Alaska's experience with oil wealth shows how states with large single-resource revenue can use broad-based taxes to increase resilience. North Slope oil is a finite resource, and as oil production declines, the state cannot continue to fund broad-based services primarily from oil; Alaska must diversify future revenue sources. Knapp presents arguments for broad-based taxes, including diversifying revenues to reduce government revenue volatility and vulnerability to revenue shortfalls. This provides an alternative to resource-revenue dependency that can make public revenues more volatile and unpredictable. The absence of broad-based taxes increases fiscal dependence on the industry, which may slow future industry growth. Dependence on resource revenues may also discourage investment by firms that fear abrupt and potentially high future taxes when resource revenues cease (Knapp, 2015). These points illustrate ways to overcome risks of relying on the Arctic's narrow resource bases.

Murmansk, Russia

The Russian North faces many challenges to economic sustainability (Suutarinen, 2013; Suopajärvi et al., 2016; Tykkylainen, 2008). Riabova and Didyk (2014) explored the Kirovsk and Apatity municipalities in Murmansk and two mining companies—Apatity and North-Western Phosphorous Company Ltd.—and the social licensing of mining companies in Russia. Results showed that municipal officials, politicians, and mining company representatives were unfamiliar with the concept of social licensing, with implications for the adoption of sustainable mining practices. In this case, public hearings were not used for decisions on the content of agreements. The lack of public hearings, institutional and organizational weaknesses of civil society, and low levels of cross-community social capital presented barriers for community participation to influence mining activities.

Canadian North and Little Cornwallis Island, Nunavut

In the case of the Polaris mine (1982–2002) operated by Cominco on Little Cornwallis Island, Nunavut, Green (2013) explores the Canadian government's shift away from supporting mining developments in the late 1970s to early 1980s, Inuit employment in the mining industry, and the difficulties of Inuit from Resolute Bay in obtaining employment. Mining development did not include local Inuit in the decision-making process, nor did Cominco prioritize hiring local people. Furthermore, the government did not require, regulate, or monitor Inuit employment. Most labor and infrastructure were imported, and the ore product was exported, impacting the local community and economy (Green, 2013). This example relates to several common Arctic challenges, including leakage of economic benefits and lack of local ownership and control.

Regarding limited benefits received by local residents in the Canadian North, Leadbeater (2009) refers to a new crisis of hinterland economic development. He argues that hinterland economic conditions in Canada have changed fundamentally and adversely since the 1970s, particularly in single-industry mining communities, where population loss can be linked to the export of economic rents and profits. Mining communities and labor have received diminished benefits from resource development as corporations externalize their social costs at the expense of local communities and labor. Redistribution of power toward communities will be needed to mitigate and counteract this trend, with a growing role for the public sector, cooperatives, and community organizations.

Kiruna iron ore mine, Northern Sweden

An extreme and perhaps unique case of negative mining impacts is illustrated by the Swedish Luossavaara-Kiirunavaara AB (LKAB) mine. Located in Kiruna (population ~17,000) in northern Sweden, this mine is the world's largest, most modern underground iron ore mine. Mining is a major source of national income and employment in Kiruna. However, its operation is undercutting the town center and forcing relocation of the town (Nilsson, 2010). This case indicates how a region responds to challenges presented by resource extraction. While Kiruna's relocation raises mixed feelings and concerns about social and cultural wellbeing, relocation and continued mining preserve important economic benefits and local hiring.

Dawson City, Yukon Territory

Population change and out-migration challenge economic sustainability when population loss leaves remote northern places too small to perform economically. However, in-migration can play an important role in economic transition. In the Canadian North, Steel and Mitchell (2017) investigate how in-migrants contribute to development goals by enhancing place identity, mobilizing local labor, and building local factor capacities in historically mine-dependent places.

Fly-in-fly-out in North America

Fly-in-fly-out (FIFO) work settings may provide another solution to volatile resource economies. FIFO represents a "no-town" model, which has become the new standard for remotely located resource development and has consequences for the sustainability of communities near resource development and for distant communities from where labor is drawn (Storey, 2010; Storey & Hall, 2017). Storey argues that the scale of activity and proximity to existing communities are factors influencing the effects of FIFO on host region communities. While FIFO operations provide communities with employment diversification and economic development opportunities, it may prove destructive if it leads to infrastructure and service demands that communities cannot meet. As for the local area where resource development occurs, the economic effects of FIFO resource development will depend on local measures to control resource decision making.

Yukon and mine closures

Other challenges exist when resource production ends. Petrov (2010) examined the economic effects of mine closures and post-mine demographic shifts in Yukon during the late 1990s economic crisis when the Faro and Beaver Creek extraction sites closed and fiscal instability and transfer dependency ensued. Using an input–output and demo-economic model to simulate direct, indirect, and induced effects of mass mine closures and subsequent population change, he found employment losses in the resource and high-tech and high-salary industries, which were most favorable for the region's future. The study suggested that policies of alienation and exploitation were applied by southern actors, thereby reinforcing boom–bust cycles that worked against achieving long-term sustainability, but the decoupled character of the staple economy mitigated the effects of mine closures.

Red Dog, Alaska

In some cases, locals benefit from resource development, often because favorable institutional arrangements allow local ownership and control. The Red Dog open pit zinc mine in northwest Alaska is located above the Arctic Circle and began operating in 1989. One of the largest zinc mines worldwide, it was developed between a resource company and the NANA Regional Corporation, an Alaska Native corporation owned by the Inupiat. In 1982, NANA signed the Development and Operating Agreement for Red Dog that gave Cominco (now Teck) the right to mine. The mine is an important employment source for the predominantly native local community and a source of revenue for NANA and other Alaska Native corporations. Indigenous ownership and control over resource extraction and economic benefits ensures accrual to surrounding communities

Loeffler (2015) evaluates the Red Dog mine's effects on 11 remote, predominantly Iñupiat, Native communities regarding jobs and income, governance, education, and subsistence activities. He suggests that significant positive community effects can be attributed to institutional relationships between regional organizations. Increased economic and employment opportunities locally and an associated increase in median household incomes result from the high local hire rate and benefit sharing with native corporations throughout Alaska. Concerns persist, however, related to subsistence activities, such as local displacement of caribou, whales, and Dolly Varden char, despite efforts to maintain and protect these activities and species. Such concerns require scientific monitoring and working with locals and their traditional and local knowledge. This case provides lessons for achieving positive community effects, such as increased material wellbeing and education, for other rural communities.

Canadian Arctic diamond mining

In some cases, Indigenous people benefit from mining via impact assessment processes and negotiated agreements. In a case study of the environmental and socioeconomic impacts in Canadian Arctic diamond mining, Couch (2002) examined Broken Hill Proprietary's diamond project in the Northwest Territories, illustrating how a two-step process—environmental IA review followed by negotiated agreements—enabled different parties to solve multiple interdependent policy, environmental, social, legal, administrative, and economic issues in a remote, sparsely populated region. This exemplifies how a local Indigenous population benefited economically from a mining project without negative consequences to traditional lifestyles.

Northern tourism

Other examples illustrating important elements of economic sustainability are found in tourism (Eligh, Welford, & Ytterhus, 2002; Grimsrud, 2017), the social economy (Southcott, Walker, Wilman, Spavor, & MacKenzie, 2010), and mixed economy settings (Holen et al., 2015; Poppel, 2006). Economic diversification through tourism can achieve a higher level of economic development, as in Churchill, Manitoba (Newton, Fast, & Henley, 2002), where tourism mobilizes resources and broad stakeholder cooperation. Merits include the relatively small capital needed and a non-consumptive approach.

Tourism is a successful non-resource dependent alternative in many Arctic locales. In Iceland, it has contributed to economic recovery and stabilization following the 2008 economic sector collapse (Wade & Sigurgeirsdottir, 2012; Iceland Chamber of Commerce, 2017). As the industry booms, however, challenges include effects on the environment, housing market, prices, and the exchange rate (Iceland Chamber of Commerce, 2017, p. 53). Cunningham, Huijbens, and Wering (2012) examine whether whale watching in Iceland offers an alternative economy for whaling and fishing communities in an era of conflict over sustainable resource use. They focus on whether nations practicing whaling can simultaneously promote whale watching, concluding that a transition to whale watching may achieve more sustainable marine resource use and a more sustainable economic future. Similarly, on Svalbard, Kaltenborn (2000) finds that while Arctic tourism represents potential for growth and earnings, further regulation is needed to achieve long-term sustainable development (Kaltenborn, 2000). Calls for tourism regulation is a common theme in the literature and raises important questions about the need to establish viable non-resource dependent alternatives for Arctic economies.

The non-profit Northern co-operative experience

The non-profit Northern co-operative experience is important for local communities. It provides an alternative solution to provide for the material wellbeing of northern residents. MacPherson (2009) show how cooperatives have displayed entrepreneurial capacities on local and regional levels, contributing to financial, human, and social capital. They also contribute to empowering Indigenous peoples. For example, Alaska's non-profit sector fills a critical gap between government services and community needs and drives change and innovation (McMillian, Wolf, & Cutting, 2015). Similarly, non-profits and co-operative organizations contribute importantly to the social economy in Northern Canada (Southcott & Walker, 2009).

Mixed economy

Mixed economies have components of market exchange, subsistence activities, culture, and tradition (BurnSilver, Magdanz, Stotts, Berman, & Kofinas, 2016; Holen et al., 2015). Evidence suggests that transition to a mixed economy strengthens the viability of northern economies. Myers (1996) argued that non-renewable resource developments had not ensured adequate or stable employment for northern peoples or for imported workers in the Canadian North and local communities adapted by developing a mixed household economy. Despite predictions of the demise of the traditional economy, this sector persists for economic and cultural reasons. As the mixed economy requires cash income to help support harvesting activities, more must be done to increase employment opportunities in the North, which will provide for a more sustainable economic future.

West-Greenland, cod-to-shrimp fishing

Contrasting paths of two Greenlandic towns, Sisimut and Paamiut, in the transittion from cod to shrimp fishing during the 1960s to 1990s elucidate the importance of initial conditions and path-dependency in determining a place's economic sustainability. The government chose Paamiut, a small town in southwest Greenland, to be a model of modernization and a major growth center in the 1960s. The invested in fisheries and built a large fish-processing plant. However, when the cod stock declined, the economy deteriorated, and the town stagnated. In Sisimut, south of Disco Bay, fisheries have also been important since the 1960s. But due to Sisimiut's proximity to the first concentrations of shrimp, it is the largest business center north of Nuuk today. The first to exploit this new resource and build harvesting capacity for it, Sisimiut's adaptive capacity allowed it to exploit this advantage. It had a more diverse economy and wider range of workforce skills than Paamitut, providing residents with an adaptive advantage. Additionally, Sisimiut had more social capital in the form of networks, enterprising spirit, and social cohesion, whereas Paamiut's fishing devleopment had been driven by external initiatives. Sisimiut showed resilience to external challenges, whereas Paamiut did not (Hamilton, Brown, & Rasmussen, 2003, p. 274). In this case, outcomes are related to initial conditions, which include natural capital and factors mediating between initial conditions and outcomes, such as physical, human, and social capital. These divergent cases illustrate northern economic challenges, including dependence on external decision making.

Pond Inlet, Nunavut

Pond Inlet (population ~1,600) is a predominantly Inuit community on northern Baffin Island with a service-based economy where government is the largest employer. Myers and Forrest (2000) explored local economic changes there from 1987–1997, describing the changing economic sectors and factors. Many obstacles to development identified in 1987 continued in 1997, including lack of infrastructure and insufficient local control and decision making. Significant changes in employment occurred, but challenges imposed by outside agencies remained, including availability of financial and business-support services and infrastructure, utility cost pricing, and freight rates. This case reveals challenges faced by most Arctic communities today, including high operating costs, leakage of money to southern goods and services, and lack of local control over and input into development policy.

Lessons from economics

In underdeveloped economies such as those in the Arctic, achieving sustainability can be difficult because of the economy's inability to provide for the wellbeing of local populations. Resource-based economies experience boom–bust cycles in which a community's wellbeing depends on the ups and downs of an industry's market. The economic health of a government-supported economy is determined by decisions made in state capitals. The ability of an economy to provide for its residents might also be limited by population growth.

The resource curse hypothesis—that natural resource production will make a region worse off—dominates discussions of resource development and sustainability. This seemingly counter-intuitive finding, that something that brings wealth may be a curse and not a blessing, may hinder the creation of a sustainable economy. Explanations include many of the characteristics of the Arctic economy: unbalanced growth, negative effects on existing industries, forgetting to save, and external imposition of costs without benefits (Frankel, 2010). However, the resource curse is not inevitable (Humphrey, Sachs, & Stiglitz, 2007): institutions matter to the outcome of resource development for local economies. Those places with good institutions use resource production to improve development outcomes (O'Faircheallaigh, 2017). Institutions are thus central to the creation of sustainable economies.

Structural change occurs when economic conditions change because of growth. Thompson (1968) suggested that such changes may make urban economies more resilient when important industries decline. He argued that when economies grew beyond a critical size, structural characteristics such as diversification, political power, fixed investments, local market growth, and economic leadership may result in irreversible growth. Huskey (2011) extended this understanding of internal change to explain resilience of Alaskan communities. While significant differences exist between large metropolitan regions and Northern economies, it is possible that changes in the internal structure of an economy because of past economic growth may help communities respond to challenges. Past economic growth may create conditions for a sustainable economy.

Finally, literature on the Green Economy provides rules to guide an economically sustainable future. This literature recognizes that the economy and the environment are linked, and the environmental costs of increasing human wellbeing should be minimized (Pezzy & Toman, 2002a). 'Sustainable economics' is a term with competing definitions, but all recognize that natural resources are limited and overused. If natural and human-made capital are good substitutes, wellbeing can be maintained as long as natural capital is replaced by man-made capital. If they aren't good substitutes, the more difficult problem of using resources to balance consumption appears (Pezzy & Toman, 2002b). These rules suggest that saving is significant for creating sustainable Arctic economies.

References cited

Aarsæther, N., Riabova, L., & Bærenholdt, J. O. (2004). Community viability. In J. N. Larsen (Ed.), *Arctic human development report* (pp. 139–154). Retrieved from www.svs.is/static/files/images/pdf_files/ahdr/English_version/AHDR_chp_8.pdf

Andersen, T. M. (2015). *The Greenlandic economy—structure and prospects. Economics working papers*. Denmark: Department of Economics and Business Economics, Aarhus University.

Bjørst, L. R. (2017). Uranium: The road to "economic self-sustainability for Greenland"? Changing uranium positions in Greenlandic politics. In G. Fondahl & G. N. Wilson (Eds.), *Northern sustainabilities. Understanding and addressing change in the circumpolar* (pp. 25–34). Cham, Switzerland: Springer.

Bolton, R. (1992). 'Place prosperity vs people prosperity' revisited: An old issue with a new angle. *Urban Studies, 29*(2), 185–203.

Bone, R., & Bone, R. (2009). Environmental impact of resource projects. In *The Canadian North. Issues and challenges*. Oxford: Oxford University Press.

Brundtland, G. H. (1987). *World commission on environment and development*. Oxford: Oxford University Press.

BurnSilver, S., Magdanz, J., Stotts, R., Berman, M., & Kofinas, G. (2016). Are mixed economies persistent or transitional? Evidence using social networks from Arctic Alaska. *American Anthropologist, 118*(1), 121–129.

Couch, J. W. (2002). Strategic resolution of policy, environmental and socioeconomic impacts in Canadian Arctic diamond mining: BHP's NWT diamond project. Impact assessment and project. *Appraisal, 20*(4), 265–278.

Cunningham, P. A., Huijbens, E. H., & Wering, S. L. (2012). From whaling to whale watching: Examining sustainability and cultural rhetoric. *Journal of Sustainable Tourism, 20*(1), 143–161.

Duhaime, G., Caron, A., Lévesque, S., Lemelin, A., Mäenpää, I., Nigai, O., & Robichaud, V. (2017). Social and economic inequalities in the Circumpolar Arctic. In S. Glomsrod, G. Duhaime, & J. Aslaksen (Eds.), *The economy of the north* (pp. 11–23). Retrieved from www.ssb.no/en/natur-og-miljo/artikler-og-publikasjoner/the-economy-of-the-north-2015

Duhaime, G., & Larsen, J. N. (2004). Economic systems. In J. N. Larsen (Ed.), *Arctic human development report* (pp. 69–84). Retrieved from www.svs.is/en/projects/ahdr-and-asi-secretariat/ahdr-chapters

Eira, I. M. G., Magga, O. H., Bongo, M. P., Sara, M. N., Mathiesen, S. D., & Oskal, A. (2008). *The challenges of Arctic reindeer herding: The interface between reindeer herders' traditional knowledge and modern understanding of the ecology, economy, sociology and management of Sámi reindeer herding*. Retrieved from http://library.arcticportal.org/550/1/Eira_127801.pdf

Eligh, J., Welford, R., & Ytterhus, B. (2002). The production of sustainable tourism: Concepts and examples from Norway. *Sustainable Development, 10*(4), 223–234.

Fidler, C. (2009). Increasing the sustainability of a resource development: Aboriginal engagement and negotiated agreements. *Environment, Development and Sustainability, 12*(2), 233–244.

Ford, J. D., Smit, B., & Wandel, J. (2006). Vulnerability to climate change in the Arctic: A case study from Arctic. *Global Environmental Change, 16*, 145–160.

Ford, J. D., Smit, B., Wandel, J., Allurut, M., Shappa, K., Ittusarjut, H., & Qrunnut, K. (2008). Climate change in the Arctic: Current and future vulnerability in two Inuit communities in Canada. *The Geographical Journal, 174*(1), 45–62.

Fountain, H., & Friedman, L. (2017, December 20). Drilling in Arctic refuge gets green light. What's next? *New York Time,* Section A, p. 21.

Fox, K., & Kumar, K. (1965). The functional economic area: Delineation and implications for economic analysis and policy. *Papers in Regional Science, 15*(1), 57–85.

Frankel, J. (2010). *The natural resource curse: A survey.* [NBER Working Paper No. 15836]. National Bureau of Economic Research, Cambridge, MA.

Glomsrød, S., Mäenpää, I., Lindholt, L., McDonald, H., & Goldsmith, S. (2017). Arctic economies within the Arctic nations. In S. Glomsrod, G. Duhaime, & J. Aslaksen (Eds.), *The economy of the north* (pp. 37–68). Retrieved from www.ssb. no/en/natur-og-miljo/artikler-og-publikasjoner/the-economy-of-the-north-2015

Green, H. (2013). State, company, and community relations at the Polaris mine (Nunavut). *Études/Inuit/Studies, 37*(2), 37–57.

Grimsrud, K. (2017). Tourism in the Arctic: Economic impacts. In S. Glomsrod, G. Duhaime, & J. Aslaksen (Eds.), *The economy of the north* (pp. 137–148). Retrieved from www.ssb.no/en/natur-og-miljo/artikler-og-publikasjoner/the-economy-of-the-north-2015

Hamilton, C. L., Brown, B. C., & Rasmussen, R. O. (2003). West Greenland's cod-to-shrimp transition: Local dimensions of climatic change. *Arctic, 56*(3), 271–282.

Hansen, A. M., & Kørnøv, L. (2010). A value-rational view of impact assessment of mega industry in a Greenland planning and policy context. *Impact Assessment and Project Appraisal, 28*(2), 135–145.

Hansen, A. M., Vanclay, F., Croal, P., & Skjervedal, A. S. H. (2016). Managing the social impacts of the rapidly-expanding extractive industries in Greenland. *Extractive Industries and Society, 3*(2016), 25–33.

Heleniak, T. (1999). Out-migration and depopulation of the Russian North in the 1990s. *Post Soviet Geography and Economics, 40*(3), 155–205.

Heleniak, T. (2014). Arctic populations and migrations. In J. N. Larsen (Ed.), *Arctic human development report.* Copenhagen: Nordisk Ministerråd.

Hermann, T. M., Sandstrom, P., Granqvist, K., D'Astous, N., Vannar, J., Asselin, H., … Cuciurean, R. (2014). Effects of mining on reindeer/caribou populations and indigenous livelihoods: Community-based monitoring by Sami reindeer herders in Sweden and First Nations in Canada. *The Polar Journal, 4*(1), 28–51.

Hill, F., & Gaddy, C. (2003). *The Siberian Curse: How communist planners left Russia out in the cold.* Washington, DC: The Brookings Institution.

Holen, D., Gerkey, D., Høydahl, E., Natcher, D., Reinhardt Nielsen, M., Poppel, B., … Aslakse, J. (2015). Interdependency of subsistence and market economies in the Arctic. In S. Glomsrod, G. Duhaime, & J. Aslaksen (Eds.), *The economy of the north 2015* (pp. 89–126). Retrieved from www.ssb.no/en/natur-og-miljo/artikler-og-publikasjoner/the-economy-of-the-north-2015

Hovelsrud, G. K., Poppel, B., Van Oort, B., & Reist, J. (2011). Arctic societies, cultures, and peoples in a changing cryosphere. In *Snow, water, ice and permafrost in the Arctic (SWIPA): Climate change and the cryosphere* (pp. 219–258). Retrieved from www.amap.no/documents/doc/snow-water-ice-and-permafrost-in-the-arctic-swipa-climate-change-and-the-cryosphere/743

Humphrey, M., Sachs, J. D., & Stiglitz, J. (2007). *Escaping the resource curse.* New York, NY: Columbia University Press.

Huskey, L. (2010). Globalization and the economies of the north. In L. Heininen & C. Southcott (Eds.), *Globalization and the circumpolar north* (pp. 57–90). Fairbanks: University of Alaska Press.

Huskey, L. (2011). Resilience in remote economies: External challenges and internal economic structure. *The Journal of Contemporary Issues in Business and Government, 17*(1).

Huskey, L. (2017). An Arctic development strategy? The North Slope Inupiat and the resource curse. *Canadian Journal of Development Studies, 39*, 89–100.

Huskey, L., Maenpaa, I., & Pelyasov, A. (2014). Economic systems. In J. N. Larsen (Ed.), *Arctic human development report.* Copenhagen: Nordisk Ministerråd.

Huskey, L., & Southcott, C. (2016). That's where my money goes: Resource production and financial flows in the Yukon economy. *The Polar Journal, 6*(1), 11–29.

Huskey, L., & Taylor, A. (2016). The dynamic history of government settlements at the edge. In A. Taylor, D. Carson, P. Ensign, L. Huskey, R. O. Rasmussen, & G. Saxinger (Eds.), *Settlements at the edge, remote human settlements in developed nations* (pp. 25–48). Cheltenham: Edward Elgar Publishing.

Iceland Chamber of Commerce. (2017). *The Icelandic economy. Current state, recent developments and future outlook* (19th ed.). Reykjavik: Icelandic Chamber of Commerce.

Kaltenborn, B. P. (2000). Arctic–Alpine environments and tourism: Can sustainability be planned? *Mountain Research and Development, 20*(1), 28–31.

Knapp, G. (2015). Resource revenues and fiscal sustainability. *Economic Development Journal, 14*(2), 15–22.

Knapp, G., & Huskey, L. (1988). Effects of transfers on remote regional economies: The transfer economy in rural Alaska. *Growth and Change, 19*(2), 25–39.

Kruse, J. (2010). Sustainability from a local point of view: Alaska's north slope and oil development. In G. Winther (Ed.), *Political Economy of northern regional development* (pp. 55–72). Retrieved from www.norden.org/en/publications/publikationer/2010-521

Larsen, J. N. (2004). External dependency in Greenland: Implications for growth and instability. *Proceedings of the second northern research forum. Stefansson Arctic institute: Akureyri* (J. H. Ingimundarson & A. Golovnov, Eds.). Presented at the Velikiy Novgorod, Russia. Velikiy Novgorod, Russia.

Larsen, J. N., & Huskey, L. (2010). Material wellbeing in the Arctic. In J. N. Larsen, P. Schweitzer, & G. Fondahl (Eds.), *Arctic social indicators.* Copenhagen: Nordisk Ministerråd.

Larsen, J. N., & Huskey, L. (2015). The Arctic economy in a global context. In B. Evengard, J. N. Larsen, & Ø. Paasche (Eds.), *The new Arctic* (pp. 159–174). London: Springer.

Leadbeater, D., & Pallagst, K. (2009). Single-industry resource communities, 'Shrinking,' and the new crisis of hinterland economic development. In *The future of shrinking cities: Problems, patterns and strategies of urban transformation in a global context*. Berkeley: Institute of Urban and Regional Planning, University of California Berkeley.

Larsen, J. N., Anisimov, O. A., Constable, A., Hollowed, A. B., Maynard, N., Prestrud, P., Prowse, T. D., & Stone, J.M.R. (2014). Polar regions. In V.R. Barros, C.B. Field, D.J. Dokken, M.D. Mastrandrea, K.J. Mach, T.E. Bilir, M. Chatterjee, K.L. Ebi, Y.O. Estrada, R.C. Genova, B. Girma, E.S. Kissel, A.N. Levy, S. MacCracken, P.R. Mastrandrea, & L.L. White (Eds.), *Climate Change 2014: Impacts, adaptation, and vulnerability. Part B: Regional aspects. Contribution of Working Group II to the Fifth Assessment Report of the Intergovernmental Panel on Climate Change*. Cambridge and New York: Cambridge University Press, pp. 1567–1612.

Loeffler, B. (2015). Mining and sustainable communities: A case study of the Red Dog Mine. *Economic Development Journal, 14*(2), 23–31.

MacPherson, I. (2009). Beyond their most obvious face: The reach of cooperatives in the Canadian north. In C. Southcott (Ed), *Northern communities working together: The social economy of Canada's north*. Toronto: University of Toronto.

McMillian, D., Wolf, L., & Cutting, A. (2015). Alaska's nonprofit sector. *Economic Development Journal, 14*(2), 34–41.

Myers, H. (1996). Neither boom nor bust: The renewable resource economy may be the best long-term hope for northern economies. *Alternatives Journal, 22*(4), 18–23.

Myers, H. (2000). Options for appropriate development in Nunavut communities. *Etudes/Inuit/Studies, 24*(1), 25–40.

Myers, H., & Forrest, S. (2000). Making change: Economic development in Pond Inlet, 1987–1997. *Arctic, 53*(2), 134–145.

Newton, S. T., Fast, H., & Henley, T. (2002). Sustainable development for Canada's Arctic and Subarctic communities: A back-casting approach to Churchill, Manitoba. *Arctic, 55*(3), 281–290.

Nielsen, S. B. (2013). *Exploitation of natural resources and the public sector in Greenland*. [Background paper for the Committee for Greenlandic Mineral Resources to the Benefit of Society]. Retrieved from http://openarchive.cbs. dk/bitstream/handle/10398/9050/Soren_Bo_Nielsen_Exploitation_of_natural_ resources_and_the_public_sector_in_Greenland.pdf

Nilsson, B. (2010). Ideology, environment and forced relocation: Kiruna—a town on the move. *European Urban and Regional Studies, 17*(4), 433–442.

Nuttall, M. (2012). Imagining and governing the Greenlandic resource frontier. *The Polar Journal, 2*(1), 113–124.

Nuttall, M. (2013). Zero-tolerance, uranium and Greenland's mining future. *The Polar Journal, 3*(2), 368–383.

O'Faircheallaigh, C. (2017). Using revenues from Indigenous impact and benefit agreements: Building theoretical insights. *Canadian Journal of Development Studies, 39*(1), 101–118.

Parlee, B. L. (2015). Avoiding the resource curse: Indigenous communities and Canada's oil sands. *World Development, 74*, 425–436.

Petrov, A. N. (2007). A look beyond metropolis: Exploring creative class in the Canadian periphery. *Canadian Journal of Regional Science/Revue Canadienne Des Sciences Régionales*, 451–474.

Petrov, A. N. (2010). Post-staple bust: Modeling economic effects of mine closures and post-mine demographic shifts in an arctic economy (Yukon). *Polar Geography, 33*, 1–2.

Petrov, A. N. (2016). Exploring the Arctic's other economies: Knowledge, creativity and the new frontier. *The Polar Journal, 6*(1), 30–50.

Pezzey, J., & Toman, M. (2002a). Progress and problems in the economics of sustainability. In T. Tietenberg & H. Folmer (Eds.), *International yearbook of environmental and resource economics 2002/2003*. Cheltenham, UK: Edward Elgar.

Pezzey, J., & Toman, M. (2002b). *The economics of sustainability: A review of journal articles*. Discussion Paper 02–03. Resources for the Future, Washington, DC.

Poppel, B. (2006). Interdependency of subsistence and market economies in the Arctic. In S. Glomsrød, G. Duhaime, & J. Aslaksen (Eds.), *The economy of the north*. Retrieved from www.ssb.no/english/subjects/00/00/30/sa84_en/kap5.pdf

Poppel, B. (2018). Arctic oil & gas development: The case of Greenland. In L. Heininen & H. Exner-Pirot (Eds.), *Arctic yearbook 2018: Arctic development in theory and in practice*. Retrieved from https://arcticyearbook.com/images/yearbook/2018/Scholarly_Papers/19_AY2018_Poppel.pdf

Prowse, T., Furgal, C., Chouinard, R., Melling, H., Milburn, D., & Smith, S. L. (2009). Implications of climate change for economic development in Northern Canada: Energy, resource, and transportation sectors. *AMBIO, 38*(5), 272–282.

Rasmussen, R. O. (2011). A question of accessibility. In R. O. Rasmussen (Ed.), *Megatrends* (pp. 168–185). Copenhagen: Nordic Council of Ministers.

Rasmussen, R. O., Roto, J., & Hamilton, L. (2014). West-Nordic region. In J. N. Larsen & A. N. Petrov (Eds.), *Arctic social indicators (ASI-II): Implementation* (pp. 139–182). Copenhagen: Nordic Council of Ministers.

Riabova, L., & Didyk, V. (2014). Social license to operate for mining companies in the Russian Arctic: Two cases in the Murmansk region. In *Arctic yearbook 2014*. Retrieved from https://arcticyearbook.com/images/yearbook/2014/Briefing_Notes/5.Riabova.pdf

Sellheim, N. (2015). The goals of the EU seal products trade regulation: From effectiveness to consequence. *Polar Record, 51*(3), 274–289.

Southcott, C., & Walker, V. (2009). A portrait of the social economy in Northern Canada. *The Northern Review, 30*, 13–36.

Southcott, C., Walker, V., Wilman, J., Spavor, C., & MacKenzie, K. (2010). *The Social Economy and Nunavut: Barriers and Opportunities*, Research Report Number 1: SERNNoCa Research Report Series. Whitehorse: Northern Research Institute. Accessed at http://yukonresearch.yukoncollege.yk.ca/frontier/files/sernnoca/PortraitureReportforNunavutv.pdf.

Steel, C. E., & Mitchell, C. J. A. (2017). Economic transition in the Canadian north: Is migrant-induced, neo-endogenous development playing a role? *The Journal of Rural and Community Development, 12*(1), 55–74.

Storey, K. (2010). Fly-in/Fly-out: Implications for community sustainability. *Sustainability, 2*(5), 1161–1181.

Storey, K., & Hall, H. (2017). Dependence at a distance: Labour mobility and the evolution of the single industry town. *The Canadian Geographer, 62*(2), 225–237.

Sugden, D. (1982). *Arctic and Antarctic: A modern geographical synthesis*. Totowa, NJ: Barnes and Noble.

Suopajärvi, L., Poelzer, G., Ejdemo, T., Klyuchnikova, E., Korchak, E., & Nygaard, V. (2016). Social sustainability in northern mining communities: A study of the European North and Northwest Russia. *Resources Policy, 47*, 61–68.

Suutarinen, T. (2013). Socioeconomic restructuring of a peripheral mining community in the Russian North. *Polar Geography, 36*(4), 323–347.

Thompson, W. R. (1968). *A Preface to Urban Economics*. London: RFF Press.

Trump, B. D., Kadenic, M., & Linkov, I. (2018). A sustainable Arctic: Making hard decisions. *Arctic, Antarctic, and Alpine Research, 50*(1), 10.

Tussing, A. (1984). Alaska's petroleum-based economy. In T. Morehouse (Ed.), *Alaska resource development, issues of the 1980s* (pp. 51–78). Boulder: Westview Press.

Tykkyläinen, M. (2008). The future of the 'boom and bust' landscape in the Russian North. In V. Rautio & M. Tykkyläinen (Eds.), *Russia's northern regions on the edge: Communities, industries, and populations from Murmansk to Magadan* (pp. 163–182). Helsinki: University of Helsinki.

Wade, R. H., & Sigurgeirsdottir, S. (2012). Iceland's rise, fall, stabilization and beyond. *Cambridge Journal of Economics, 36*(1), 127–144.

Wilson, E. (2016). Negotiating uncertainty: Corporate responsibility and Greenland's energy future. *Energy Research & Social Science, 16*, 69–77.

3 Culture and sustainability

Susanna Gartler, Vera Kuklina,
and Peter Schweitzer

Introduction

A search of popular academic databases, such as GoogleScholar, Articles-Plus, and Web of Science, reveals that combinations of the keywords "Arctic, sustainable, and culture" are very common. Thousands of publications use these terms. Moreover, numerous studies focus on similar issues without using the keywords "culture, sustainable, or sustainability," demonstrating the popularity of the topics and the urgency to examine the nexus of culture and sustainability.

'Cultural sustainability' and 'sustainable cultures' can mean different things. Cultural sustainability can refer to cultural aspects of a society that relate to ecological welfare, or it can refer to the longevity and vitality of a thriving culture or society. Sustainable culture may also refer to cultures that value ecological welfare or that are vital and resilient and will likely thrive for a long time. We pursue two possible constellations: sustainable cultures, focusing on prospects of cultural viability in Arctic contexts and cultural sustainability, which explores the role of culture(s) in broader sustainability contexts.

A model by Soini and Dessein (2016) differentiates how culture and sustainability interact conceptually. Depending on which option is applied, the relationship between culture and society, culture and nature, culture and development, and the definition of culture itself vary (Figure 3.1).

Cultural sustainability: a fourth pillar?

Soini and Dessein (2016) explain that the culture-in-sustainability framework corresponds to culture as a fourth pillar of sustainability. Acknowledging the "limitations of the 'pillar approach' to sustainability" (Soini & Dessein, 2016, p. 2), they take it as a starting point because it is commonly used. In much of the literature on cultural sustainability, Hawkes

	Culture in sustainability	Culture for sustainability	Culture as sustainability
Definition of culture	Culture as capital	Culture as a way of life	Culture as a semiosis
Culture and development	Culture as an achievement in development	Culture as a resource and condition for development	Development as a cultural process
Value of culture	Intrinsic	Instrumental and intrinsic	Embedded
Culture and society	Complementing	Affording	Transforming
Culture and nature	Human perspective on nature	Interaction of culture and nature	Nature constituent of culture
Policy sectors	Cultural policies	All policies	New policies
Modes of governance	Hierarchical governance, first order	Co-governance, second order	Self-governance, meta-governance
Research approach	Mainly mono- and multidisciplinary	Mainly multi- and interdisciplinary	Mainly inter- and transdisciplinary

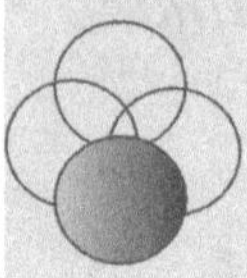

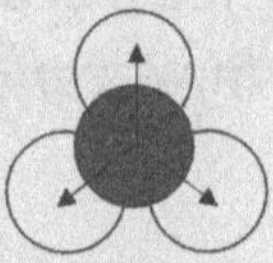

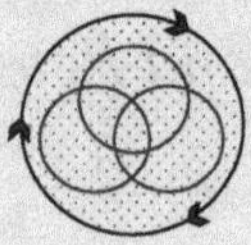

Figure 3.1 Conceptual representation of understandings of sustainability (after Soini & Dessein, 2016)

(2001) is credited with introducing culture as a fourth pillar of sustainability (see Baycan & Girard, 2013; Birkeland, 2008; Loach, Rowley, & Griffiths, 2016; Polistina, 2009; Soini & Birkeland, 2014; Stylianou-Lambert, Boukas, & Christodoulou-Yerali, 2014), but the conceptualization originated from Yencken and Wilkinson (2001). Prior to that, Throsby (1995) introduced the idea of 'culturally sustainable development' to link culture and economy and to add to the Brundtland definition that links only economy and ecology.

According to the Brundtland report, sustainable development emphasizes two key concepts: the needs of the poor and "the idea of limitations imposed by the state of technology and social organization on the environment's ability to meet present and future needs" (Brundtland, 1987, p. 43). This definition identifies the importance of the needs of the disadvantaged and the ability of the environment to meet human needs. Needs, of course, entail culturally specific dimensions. Human–environmental relations are likewise determined by the ideational, moral, and symbolic realms of a community at a certain time. Therefore, cultural aspects are an integral sub-text of this definition of sustainable development.

Recently, publications increasingly address cultural sustainability (Asikainen, Brites, Plebańczyk, Mijatović, & Soini, 2017; Soini & Birkeland, 2014). Soini and Birkeland (2014) explore the scientific discourse on cultural sustainability, noting the elusiveness and multiple interpretations of the concept, which allows diverse stakeholders to connect with it (Soini & Birkeland, 2014) because the deployment of cultural sustainability is always linked to a social context. Their analysis reveals multiple meanings conveyed through narratives that generate ideas or categories about the concept. Each theme is associated with a slightly different understanding of culture, where culture is something inherited by the next generation, a resource for economic vitality and local development or a 'spatial and temporal coherence of an area' (Soini & Birkeland, 2014, p. 218). Multiple narratives show that a clear distinction between different pillars of sustainability is hard to achieve.

Cultural sustainability is also sometimes framed as part of social sustainability (Soini & Birkeland, 2014; Myllyviita et al., 2014); Low (2008) considers social sustainability as a subset of cultural sustainability. However, Abu-Lughod (1991) describes how cultural difference is often invoked in cases of conflict, migrants, or people of color, whereas when the problem pertains to white people, the term used is 'society.' How can cultural sustainability, then, be meaningfully differentiated from social sustainability? Dessein, Soini, Fairclough, and Horlings (2015) assume that "culture and society have to some degree an iterative and reciprocal relationship, in which culture constructs society but society also shapes culture" (Dessein et al., 2015, p. 25). But as the model by Soini and Dessein (2016) shows, how this relationship is understood depends on the framing of the relationship between sustainability and culture.

Examining Arctic sustainability issues, Petrov et al. (2016) discuss "economics, ecology and equity" (2016, p. 172, see also Petrov et al., 2017, p. 3). A difference between cultural sustainability and other sustainability concepts seems that the former invokes the notion of culture, whereas the latter doesn't. The variety of notions of cultural sustainability that we encounter calls for a closer look at its components, providing both critique and ways of operationalizing culture and sustainability.

Operationalization and critique of culture and sustainability

Culture

Studies of culture, especially in cultural and social anthropology, are defined by this keyword, and many authors treat 'culture' as anthropology's defining theoretical contribution (Hann, 2002; Sahlins, 1999; Trouillot, 2002). Simultaneously, criticisms of the concept, notably Abu-Lughod (1991), Gupta and Ferguson (1992), Stolcke (1995), and Dombrowski (2001), suggest practices of 'othering' and processes of exclusion and inclusion are enabled through 'culture.' Ingold (2018) calls culture one of the "two inner demons" (p. 123) of anthropology (the other being race). Both, he explains, have philosophically presided over homicide, ethnic cleansing, and genocide, and inheritance and essentialism are identified as the foundational principles of these concepts, making them dangerous (Ingold, 2018, p. 123). Combined, the ideas of fixed, immutable groups and the biological transmission of cultural and/or physical traits can be lethal, as history has shown. Clammer (2016) critiques the culture within sustainability, arguing that it is capitalist culture that prevents sustainability. The term 'culture' has also become ubiquitous outside of anthropology, reflecting a generally non-critical stance toward it (Sahlins, 1999).

Yencken and Wilkinson (2001), who introduced the idea of the fourth pillar of sustainability, examine culture within sustainability by emphasizing how the environment shapes culture directly and indirectly, corresponding to the culture-for-sustainability idea in the Soini and Dessein's (2016) model. Hawkes (2001) discusses the concept of culture by explaining that sustainable culture is vital, creative, and innovative; embraces difference and the arts; and fosters belonging and a sense of identity. Focusing on culture as a value system vital to sustainability, Hawkes argues that the wish for one's own culture to inform the culture of future generations is a 'legitimate desire' (2001, p. 11).

Attempting to operationalize culture, Kivitalo et al. (2016) name a worldview, a symbolic, a reification, and an institutional dimension. These distinctions are useful, but many phenomena permeating these four dimensions—organization, for example—in which the authors' characterization of worldview and institution come together to form values, behavioral norms, and codes of conduct in businesses, administrative bodies, families, kin groups, and so on. Another dimension is the artistic realm, where symbolisms, art forms, and cultural practices also define culture.

In the North American Indigenous context, culture is increasingly important for healing inter-generational traumas and the effects of violent colonial histories (Jacob, 2013; Castelano, 2009). Culture may fill the gap left by colonialism and demise of religions: where people used to find god or communism, now they find culture. This line of reasoning ignores how colonial thought and religious systems are imposed on functioning societies, which have never been completely erased, with some exceptions. Moreover, many problems Indigenous people face today were partly caused by the imposition of Christian religions, Western thought systems, and secular or atheist education, especially in post-Soviet space.

Thinking through culture and sustainability often leads to more questions than answers. Does 'sustainable culture' mean that a culture is maintained in the same state forever? How can change be incorporated? Cultural persistence and change are not contradictory characteristics of a given culture, society, or group: continuity and constant change are always part of life. A purely static understanding contradicts omnipresent social and cultural change, whereas an understanding of 'cultures being in constant flux' does not reflect feelings of loss of culture. As Kirsch (2001) explains: "Indigenous claims about 'culture loss' pose a problem for contemporary definitions of culture as a process that continually undergoes change rather than something which can be damaged or lost" (p. 167).

The *Arctic Human Development Report* (*AHDR*) exemplifies how the non-static nature of the term culture occurs in contemporary anthropological thought:

> In order to avoid deterministic views of culture, we define [culture] as a non-static, creative process. . . . All the configurations of meanings, sounds, relations and logic change over time but what remains is their embeddedness in socioeconomic environments and systems of adaptations to the Arctic landscape.
>
> (Schweitzer & Ulturgasheva, 2014, p. 105)

Similarly, Golovnev (2012) develops the idea of 'changeability' as the main characteristic of Arctic cultures, which presents Arctic cultures as dynamic temporally (Pivneva, 2015) and spatially (Golovnev, Garin, & Kukanov, 2016). The *AHDR*'s definition accentuates fluidity and change while simultaneously employing the notion of embeddedness. Indigenous groups, not least as a consequence of multiple enforced separations (Neufeld, 2016), emphasize continuity and longevity of their ways of life (LaRocque, 2010). Adaptive capacities and resilience as part of stability have long characterized Arctic Indigenous cultures and have become more important with the rapidity and scale of current changes. Nonetheless, all Arctic cultures must be seen as dynamic processes of group identity formation.

Sustainability

Erik Swyngedouw offers a critical analysis of sustainability discourse. He remarks that it often invokes humanity as a whole, leaving aside tensions and conflicts inherent within capitalism and liberal politics. Drawing on Marxist philosopher Slavoy Žižek, he calls sustainability a "consensually established concern . . . structured around ecologies of fear that nurture a reactionary stance and urge techno-managerial forms of intervention" (2010, p. 312). Indicators of degrees of sustainability (Hawkes, 2001) could be understood as techno-managerial action instead of "asking the politically sensitive, but vital, question about the kinds of socio-environmental arrangements and assemblages we wish to produce, how this can be achieved, and what sort of environments we wish to inhabit" and sustainability the "signifier that encapsulates these post-political attempts to deal with Nature" (Swyngedouw, 2010, p. 310).

Swyngedouw further criticizes sustainability culture, suggesting that this is only a gesture by the powerful to indicate that they are taking environmental concerns seriously. Referring to Lacan's 'structure of fantasy,' he argues that the meaninglessness of the term is a canvas for projection of elite fantasies of an idealized world. He critiques the externalizing of problems associated with modern-day capitalism and liberal politics (Swyngedouw, 2010). Keeping in mind Crate's (2006) criticism of the Brundtland report as a reinforcing, dominant, Western worldview and a generic solution to particular sets of problems, sustainability can ever only be partially operationalized.

Moreover, sustainability as a concept is linked to the development of protocapitalism (Radkau, 2011). Radkau argues that before large-scale exploitation of forests, the idea of sustainability was redundant. Small-scale farmers and forest users were aware of the time needed for trees to grow to use wood resources (Radkau, 2011, p. 94); sustainability is inextricably linked to ideas about development (Petrov et al., 2017). Clammer (2016, p. 8) writes:

> At the basis of this of course lies the fundamental question: what is development or "progress" for? Is it simply about the expansion, maximization and acquisition of material resources? More technology? More "growth"? Such a view, with its emphasis almost entirely on the economic aspects of development is a highly impoverished one. It excludes not only the rights of nature and other species which share the planet, but also the aesthetic, spiritual and expressive needs of human beings.

The concept of development and its inherent ideologies of fast-paced, large-scale industrialization is criticized and rejected among many Arctic inhabitants and policymakers (Petrov et al., 2017). Is it, then, possible to employ sustainability at all? Soini and Dessein (2016, p. 2) differentiate between the sustainability and sustainable development, which are often used synonymously, and contend that

> The concept of sustainable development is often criticized of being in favor of growth, efficiency, and the increase of technology, although development can also be considered in a qualitative way. Sustainability, on the other hand can be understood not only as a universal goal to be achieved, but as a procedure or continuously evolving "imaginary world."

Some sustainability approaches focus less on its philosophical underpinnings and societal effects and instead on practical aspects, such as how to achieve and measure it (Caradonna, 2014). Caradonna (2014) asserts that the term is a way to express disagreement with environmental degradation caused by industrialization processes and "a way of acknowledging how humankind has created an [ecological] imbalance." (p. 3). Meitz and Ringhofer (2017) write that sustainability must be considered alongside resilience because "it is the combination of both concepts which allows for a long-term sustainable solution which also buffers from external shocks." (p. 159)

Spatial and temporal scales

Petrov et al. (2016) discuss themes such as policy and research implications and identify a lack of concern with spatial and temporal scales of sustainability (see also Clammer, 2016). Many spatial configurations can be identified in discussions about sustainability and may revolve around mega-cities, small towns, remote rural areas, mountains, the seaside, even sub-surface and lacustrine environments. In fact, spatiality and temporality represent two key differentiating factors. For example, an archaeological or historical case study may highlight how long-term temporal frames shaped prior resource use. However, examining temporal frames guiding the rationale behind current resource extraction reveal that industry usually favors shorter time periods and particular geographies, or spaces.

Petrov et al. (2017) note that socio-cultural wellbeing of Indigenous communities is important for Arctic sustainability alongside resource exploitation. Indigenous people, for example, may emphasize a multi-generational approach, noting that mining can be done in the future and for that reason should be done responsibly. This temporal framework is different from other understandings.

Kivitalo et al. (2016) address spatial and temporal considerations and consider sustainability in four so-called binaries: spatial, temporal, ethics, and diversity. Emphasizing space, place making, rurality, and practices of territorialization, they draw conclusions from their case study in Finland. They provide a framework for understanding the interplays of culture and sustainability in rural areas and in cities and urban dwellings. In many discussions on sustainability, topics such as food security, climate change, urbanization, rural development, gender issues, local knowledge, human–animal relations, and many others quickly arise. Thus, whereas the term is very broad and can be criticized from a Marxist philosophical viewpoint, many sustainability issues are concrete and relate to inequalities and tensions within the global market system. Space and time may serve, then, "as meta-level aspects of analysis" (Meitz & Gartler, 2017, p. 249).

Different terms, similar concerns

Some texts address boundaries in comparison to other normative concepts such as the Anthropocene (Kohn, 2015), the pluriverse (Escobar, 2011), or the ecozoic (Berry, 1999). In what Petrov et al. (2016) call Arctic sustainability research, other authors have started using other terms and concepts to describe and analyze similar phenomena, such as cultural vitality (Schweitzer, Fox, Csonka, & Kaplan, 2010) and vulnerability (Stammler-Gossman, 2010; see Haalboom & Natcher, 2013 for a critique of vulnerability and community labeling), contact with nature (Crate et al., 2010), the creative economy (Pelyasov, 2009), adaptation (Hovelsrud, White, Andrachuk, & Smit, 2010), viability (Riabova, 2010), and resilience (Snow & Ochlaski, 2018).

The ontological turn may be understood as a current response to climatic and ecological problems and, when linked to sustainability, an ethical practice that rethinks relationships toward non-humans (Kohn, 2015). A new way of thinking is necessary for a "world in jeopardy" (Kohn, 2015, p. 315), ultimately away from Cartesian thought, dichotomies, and a mechanistic understanding of the human relationship to Earth that has supported industrial exploitation and degradation (see also Birkeland, 2008).

In Russia, the study of cultural landscapes is dedicated to preserving culture and environment. With creation of the Institute of Cultural and Natural Heritage in 1992 (Vedenin, 2003) and rooted in the landscape school (Berg, 1915), current cultural landscape studies embrace the interdependence and interrelation of cultural and natural landscape components (Romanova & Nikiforova, 2015; Mikhailova, 2014; Komarova, 2014). For example, Yuri Lukin uses the metaphor of cultural landscape for examining social and cultural processes in the Arctic (Lukin, 2011, 2015a, 2015b) and Kalutskov

(2008) investigates natural components of cultural landscapes to understand interdependencies and relations within particular cultures.

In contrast, the idea of a 'circumpolar civilization' emphasizes human capacities for balanced development (Vinokurova & Yakovets, 2016). Such studies are often based on Vernadsky's (1991) conceptualization of a noösphere, a next stage of human development when humans play a crucial role in planetary-scale biological and geological processes. Similarly, the idea of the Anthropocene, when "human and non-human kinds and futures become increasingly entangled" and "ethical and political problems can no longer be treated as exclusively human problems" (Kohn, 2015, p. 312), indicates the interlinking of human behavior with Earth's future wellbeing.

The Center of Circumpolar Civilizations at the Arctic State Institute of Culture and Arts in Yakutsk studies the "sustainable development of cultures of the peoples of the Russian Federation settled within the permafrost zone" (Marfusalova, 2002; Robbek, 2000). Different features of circumpolar civilization, including subsistence cultures, identities, cultural policy, spatial images, environmental ethics, arts, and cultural landscapes, are described in the book *Arctic Culture* (Ignatieva et al., 2014).

Arctic Indigenous cultures and sustainability

What needs to be sustained? Can culture be depleted? Is culture loss or cultural change due to un-sustainability? And can we even talk about sustainable cultures or cultural sustainability when it means treating culture as a resource? Corresponding to Soini and Dessein's (2016) first model of culture-in-sustainability, in which culture is treated as capital, the notion of cultural sustainability invokes culture as a fourth pillar to achieve the goal of sustainability. Now we again must ask: Who is "we?" How does the answer allow for inequality to appear and matter, structurally?

Culture-as-resource makes sense empirically for indicating its value in Indigenous healing and reconciliation. Epistemologically, this approach is flawed because the term 'resource' is part of a Western capitalist thought system that seeks to incorporate other ontologies and ways of being. Nadasdy (2008) recognizes problems with the term 'natural resources' because conceptualizing non-humans as resources reinforces spatiotemporal assumptions of bureaucratic wildlife management. Then again, culture-as-resource allows sense to be made of culturally extractive processes, such as cultural appropriation (e.g., Germans playing Indians while believing that they are preserving North American Indigenous cultures and traditions; Lopinto, 2009). Culture is often treated as synonymous to a way of life, as in culture-for-sustainability. The third model, culture-as-sustainability, instead sees development as a cultural process in which the value of culture is embedded and corresponds to self-governance (Soini & Dessein, 2016, p. 4).

Other conceptualizations of sustainability, such as the concept of Inuit Quaujimajatuqunagit (IQ), are alternatives to westernized understandings of interacting with natural surroundings. Referred to as Inuit epistemology, IQ is knowledge that has been passed on through generations and ensures survival (Snow & Ochlaski, 2018). It corresponds most closely to culture-as-sustainability because "IQ does not adopt western language . . . but develops its own particular terminology" (Petrov et al., 2017, p. 47). Doubleday (2003) defines culture or cultural survival as an outcome and a source of sustainability, claiming that Inuit identity crucially informs the concept:

> Here it is argued that the fact that Inuit have within living memory experienced both autonomy and dependency has given Inuit an indelible sense of identity. From this fundamental sense of identity, it is then argued that this knowledge of what it means to be Inuit, what it means to be part of a community (in both the old and new meanings of community), and what is necessary to survival, further informs concepts of sustainability within Nunavut.

Other Arctic groups emphasize their sustainable cultures and actively propagate values and rules of moral conduct that indicate that humans depend on the planet, resources should be available for future generations, and we should strive to enact relationships to non-human persons accordingly. In some cases, such as the North American Northern Tutchone, this takes the form of strategic environmentalism in which the legitimacy of land claims and group cohesion are reinforced via a common narrative as stewards of the land. Sustainable culture is rooted in Northern Tutchone ontology, in which people are responsible for maintaining the "balance between the environment, the animals and our people" (Peter, Hogan and the First Nation of Na-Cho Nyäk Dun Lands and Resources Department, 2006, p. 86).

To understand the underlying principles of the Northern Tutchone ontology and how it ties into the debate about cultures for sustainability, we examine a publication by the Little Salmon-Carmacks and Selkirk First Nations (both Northern Tutchone groups, the third one being the First Nation of Na-Cho Nyäk Dun) as part of an effort to collect, preserve, and revive Doòli, spiritually driven laws governing relationships among humans, animals, and their surroundings (Natcher, 2007, p. 274). Northern Tutchone emphasize humans' responsibility as caretakers and mediators among the animal, spirit and human worlds:

> The ecosystem is out of balance. The relationship between humans and the land needs mending and the only way to do that is through Doòli.

Bringing back Doòli for the good of the land is a big job. This job begins with the Elders and they must ensure that future generations will understand and follow Doòli for all animals.

(Urquhart, 2013)

Mending the land is achieved via collection, transmission, and preservation of knowledge and approval for further action by Elders. The 'caretakers of the land' narrative occurs in many interviews and talks with community members. Northern Tutchone, other First Nations, Inuit, and Métis thus actively engage in propagating culture-for-sustainability and culture-as-sustainability narratives. Teachings emphasize respect for a sentient nature, never taking more than you need, making sure animals are doing well, giving thanks, and always using all parts of the animal. This relational worldview does not differentiate between human and non-human persons, and balance is achieved by all actors doing their part. Indigenous languages help sustain such values for future generations (Cruikshank, 2005).

Language retention

A reductionist understanding of Indigenous peoples regarding their forms of provision, such as hunting and gathering, results in little attention paid to what people consider as their culture, notably language. But language is important for sustainable or viable cultures. Moreover, if cultures or groups that openly emphasize sustainability are threatened by language loss and other assaults, such as colonialism, the very survival of that people is at stake.

Yencken and Wilkinson (2001) find that language and different terms and words shape our perceptions of our surroundings. For example, they argue that the English language is often inadequate to describe Australian flora and fauna and call for a common language that helps "value more greatly" what is around us (p. 356). They address language loss, which they frame as a "loss of culture" (p. 357), and view importing foods instead of eating what is already as a refusal to be absorbed by and connected to a place.

The second *AHDR* (*AHDR II*) notes the importance of traditional activities and forms of subsistence alongside spirituality, place attachment, landscape, arts, and sports for vital Arctic Indigenous cultures. Language vitality is an important issue because "the general tendency continues to be a reduction of the proportion of speakers of most languages of the North" (Schweitzer & Ulturgasheva, 2014, p. 114). While some progress with new laws and efforts exist and not all languages suffer from low fluency among their populations, a large proportion is under threat. Grenoble (2018, p. 351) writes that "all Arctic Indigenous communities are undergoing language shift to varying degrees and thus exhibit varying levels of endangerment, with the exception of Kalaallisut (Greenlandic), which has a firm standing as the national and official language of Greenland." Grenoble and Olsen (2014) explain how language vitality and sustainability is strongly linked to mental and physical wellbeing.

Language retention and revitalization illustrates the numerous challenges and complexities associated with assimilation policies and cultural erasure (Cruikshank, 1998). Governments and groups across the Arctic focus on effective mitigation of the effects of these processes and revitalizing languages. Factors that enable revitalization include demographic characteristics of a group, and high birth rates among Indigenous peoples and a willingness to self-identify as Indigenous in North America is a positive development in this regard (Nagel, 1996).

In Russia, alternative transfers of knowledge to younger generations are being implemented. In the "Nomad school" in the Republic of Sakha (Yakutia) (Gabysheva, 2016), culture is transferred through traditional subsistence activities. Children are educated by parents or teachers who migrate with nomadic families (Egorov & Neustroev, 2003; Vinokurova, 1997). Russian authors rarely describe other forms of transfer of culture in the Arctic in terms of sustainability. However, the journal *Arktika*, museums (Pisareva & Vinogradov, 2010; Truievtseva, 2012), and sports (Vinokurova, Zhegusov, Mestnikova, Alekseeva, & Alekseev, 2017) are mentioned as contributions to sustainable development.

Traditional land use in Russia

Most studies related to cultural aspects of sustainable development in Russia focus on Indigenous peoples. Indigenous worldviews, values, and adaptability to natural conditions are considered alternative to dominant societal values (Donahoe & Istomin, 2010; Sirina, 2008; Shapkhaev & Shapkhaev, 2005; Gudyma & Bulatov, 2002). Traditional nature management (*traditsionnoe prirodopol'zovanie*) is a specific example of Indigenous cultures in Russia. While frequently focused on preserving biodiversity and natural ecosystems, traditional nature management has become a main instrument for claims of cultural connections to land and as an opposition to industrial land use. The movement of neo-traditionalism—the revival of pre-Soviet cultural practices with modern resources (Pika, 1996)—has been an important part of sustainable development (Lamazhaa, 2010; Tiugashev, 2010).

The legal framework for this movement is explained in the federal law "On the territories of traditional management of nature" (Russian Federation, 2001). While not implemented at the federal level, studies of traditional land use by Indigenous peoples emphasize the importance of this law for sustaining traditional cultures (Klokov, 1998, 2011; Klokov, Krasovskaya, & Krasovskaya, 2001; Kriukov et al., 2014; Turaev, 1997; Mangataeva, 2000). Klokov et al. (2001) discuss Russian Old-settlers (*starozhili*), numerically small and numerically large Indigenous groups, all of whom depend on natural resource management.

Klokov (2012) critiques Russian Indigenous cultural policy on traditional land use-based economies as non-innovative and too focused on authentic traditions. To estimate reindeer herders' wellbeing, for example, he proposes analyzing retrospective herding dynamics alongside collecting statistics about today's herds and herders. The discussion about cultureas-resource may also apply here as an instrument to promote Indigenous cultures by intellectual elites (Robbek, 2010). Moving beyond the dominating extractivist discourse, some studies explore the significance of spatial human–natural relations (Ragulina, 2000).

From a legal perspective, studies of traditional lands and resource management are critiqued for their attention only to small-numbered Indigenous groups and subsistence activities (Overland, 2009). Meanwhile, Rodoman (2015) suggests that all rural dwellers in small villages lead traditional ways of life. Common problems, according to this author, include the absence of social infrastructure, over-reliance on kinship networks instead of market relations and state support, and the continuation of environmentally friendly but non-economically viable subsistence practices.

Some research addresses how small-numbered Indigenous peoples adapt to changing lives—from nomadic to settled, from rural to urban, adapting to climate change—address decreasing traditional land use practices (Tomaska, 2017; Vinokurova, 2003). In a study of urban migrants, Tomaska (2017) noted the willingness of parents to integrate children into urban societies and provide them with better educational opportunities, a process also accompanied by adopting a different, dominant culture. Studies among small-numbered Indigenous peoples have shown a high adaptive capacity to environmental changes and low adaptive capacity to changing socioeconomic conditions (Vinokurova, 2003). Some researchers may find risks of cultural loss in these trends, but future research should address specific tactics and strategies to revitalize Indigenous cultures in urban contexts.

Based on the literature, multiple factors determine sustainable cultures in Russia (and elsewhere), including use and retention of native language, subsistence, dwelling (nomadic, rural, urban, temporary), diversity of lifestyles, mobility, communication means, fate control, remoteness (center–periphery relations, local patriotism, participation in global cultures), and means of local culture transfer such as educational programs, events, films, songs, ethno-design, museums, cultural centers, and libraries.

Conclusion

Speaking about culture and sustainability means entering a normative discourse, concerning morals, ethics, values—and therefore a variety of epistemological and ontological approaches of what should be and how we should understand things. Even though these concepts are riddled with uncertainties, they are both used and discussed widely. Since culture and sustainability are terms employed for different reasons, cultural sustainability could be regarded as an empty signifier (Swyngedouw, 2010). Moreover, the seeming emptiness or excessive broadness of culture and sustainability makes their various combinations vulnerable to being hijacked by profit-seeking companies who aim to embellish themselves (Radkau, 2011) or by racist groups who argue along surprisingly similar lines as environmentalists or Indigenous groups when calling for the recognition, preservation, and revitalization of their ways of life.

'Culture' is always operationalized in particular settings and increasingly promoted by governmental agencies and non-governmental organizations as part of sustainability. This chapter emphasizes that "culture in sustainability serves as a meta-narrative that will bring together ideas and standpoints from an extensive body of academic research currently scattered among different disciplines and thematic fields" (Soini & Dessein, 2016, p. xiii). The arbitrariness regarding what is understood and discussed as part of sustainability in scientific literature points toward the weak—until recently—theoretical underpinnings of the discussion. Soini and Dessein's (2016) approach elucidates the culture–sustainability nexus.

Scholarly publications about the Arctic and beyond that use the terms 'culture' and 'sustainability' have increased recently, but it remains to be seen which paradigm will dominate future discussions. Surely, cultural sustainability is an established fourth pillar of sustainability. In the Arctic, Indigenous discourse tends to emphasize culture as a way of life and sees vital cultures and languages as sources and outcomes of sustainability (Doubleday, 2003, Grenoble & Olsen, 2014), thus locating it between the culture-for-sustainability and culture-as-sustainability approaches. Employing more of a culture-for-sustainability approach, point 36 of the United Nations Sustainable Development Goals declaration states:

> We pledge to foster inter-cultural understanding, tolerance, mutual respect and an ethic of global citizenship and shared responsibility. We acknowledge the natural and cultural diversity of the world and recognize that all cultures and civilizations can contribute to, and are crucial enablers of, sustainable development.
>
> (United Nations, 2019)

Similarly, UNESCO recognizes culture as an "enabler and a driver of the economic, social and environmental dimensions of sustainable development" (UNESCO, 2016). However, with extremely urgent problems such as global warming, youth suicide, and Indigenous language retention, all of which are exacerbated in the Arctic, necessitates a critical approach to sustainability and culture's role in the Arctic. As long as educational policies and industrial development goals are dictated by the South, change will occur slowly in the North. It is quite telling that the United Nations Sustainable Development Goals promote neo-liberal self-responsibility by adding, as a first measure to each goal, something an individual can do on a personal level to achieve sustainability instead of calling out structural inequalities that enable a system of inequity. Such an approach is not good enough: a more critical stance about capitalist culture must address important sustainability issues in the 21st century.

References cited

Abu-Lughod, L. (1991). Writing against culture. In L. Abu-Lughod & A. Appadurai (Eds.), *Recapturing anthropology: Working in the present* (pp. 137–162). Santa Fe: School of American Research Press.

Asikainen, S., Brites, C., Plebańczyk, K., Mijatović, L. R., & Soini, K. (Eds.). (2017). *Culture in sustainability towards a transdisciplinary approach.* Retrieved from https://jyx.jyu.fi/handle/123456789/56075

Baycan, T., & Girard, L. F. (2013). Culture in international sustainability practices and perspectives: The experience of 'slow city movement—Cittaslow.' In G. Young (Ed.), *The Ashgate companion to planning and culture* (pp. 273–292). Farnham: Ashgate.

Berg, L. S. (1915). Predmet i zadachi geografii. *Izvestiia RGO, 51*, 9–463.

Berry, T. (1999). *The great work: Our way into the future.* New York: Bell Tower.

Birkeland, I. (2008). Cultural sustainability: Industrialism, placelessness and the re-animation of place. *Ethics, Place, and Environment, 11*, 3–283. https://doi.org/doi:10.1080/13668790802559692

Brundtland, G. H. (1987). *Our common future: Report of the world commission on environment and development.* Oxford: Oxford University Press.

Caradonna, J. L. (2014). *Sustainability: A history.* New York, NY: Oxford University Press.

Castelano, M. B. (2009). Women's contribution to community healing. In G. G. Valaskakis, M. Stout, & E. Guimond (Eds.), *Restoring the balance* (pp. 203–236). Winnipeg, Manitoba: University of Manitoba Press.

Clammer, J. (2016). *Cultures of transition and sustainability. Culture after capitalism.* New York, NY: Palgrave Macmillan.

Crate, S. (2006). *Cows, kin and globalization: An ethnography of sustainability.* Altamira: Rowman, Alta Mira Press.

Crate, S., Forbes, B., King, L., & Kruse, J. (2010). Contact with Nature. In *Arctic Social Indicators I (ASI-I)* (pp. 109–127). Nordic Council of Ministers.

Cruikshank, J. (1998). *The social life of stories: Narrative and knowledge in the Yukon territory.* Lincoln: University of Nebraska Press.

Cruikshank, J. (2005). *Do glaciers listen? Local knowledge, colonial encounters, and social imagination.* Vancouver: UBC Press.

Dessein, J., Battaglini, E., & Horlings, L. (2016). *Cultural sustainability and regional development: Theories and practices of territorialisation, Routledge studies in culture and sustainable development.* London: Routledge.

Dessein, J., Soini, K., Fairclough, G., & Horlings, L. (2015). *Culture in, for and as sustainable development. Conclusions from the COST Action IS1007 investigating cultural sustainability.* Finland: University of Jyväskylä.

Dombrowski, K. (2001). *Against culture: Development, politics, and religion in Indian Alaska.* Lincoln: University of Nebraska Press.

Donahoe, B., & Istomin, K. V. (2010). Izmenenie praktiki regulirovaniia dostupa k prirodnym resursam u nekotoryh olenevodcheskih narodov Sibiri. Popytka teoreticheskogo obobshheniia. *Novye issledovaniia Tuvy, 4,* 11.

Doubleday, N. C. (2003). The nexus of identity, Inuit autonomy and Arctic sustainability: Learning form Nunavut community and culture. *British Journal of Canadian Studies, 16*(2), 297–308.

Egorov, V. N., & Neustroev, N. D. (2003). *Specifika dejatel'nosti malokomplektnyh kochevyh shkol v uslovijah Severa.* Moscow: Academia.

Escobar, P. (2011). Sustainability: Design for the pluriverse. *Development, 54*(2), 137–140.

Gabysheva, F. V. (2016). Obrazovatel'noe prostranstvo Arktiki: Razvitie cherez dialog i sotrudnichestvo. *Kul'tura i iskusstvo Arktiki, 2,* 40–44.

Golovnev, A. V. (2012). Etnichnost': ustoichivost' i izmenchivost' (opyt Severa). *Etnograficheskoe obozrenie, 2,* 3–12.

Golovnev, A. V., Garin, N. P., & Kukanov, A. D. (2016). *Olenevody Yamala (materialy k Atlasu kochevykh tehnologii).* Ekaterinburg: UrO RAN.

Grenoble, L. A. (2018). Arctic indigenous languages: Vitality and revitalization. In L. Hinton, L. Huss, & G. Roche (Eds.), *The Routledge handbook of language revitalization.* New York, NY: Routledge.

Grenoble, L. A., & Olsen, C. C. (2014). Language and wellbeing in the Arctic: Building Indigenous language vitality and sustainability. In *Arctic Yearbook.* Retrieved from https://arcticyearbook.com/images/yearbook/2014/Scholarly_Papers/3.Grenoble.pdf

Gudyma, A. P., & Bulatov, V. I. (2002). Social'no-filosofskie i ekologicheskie aspekty ustoichivogo razvitiia korennyh malochislennyh narodov Severa. *Ekologiia. Seriia Analiticheskih Obzorov Mirovoi Literatury, 66,* 3–109.

Gupta, A., & Ferguson, J. (1992). Beyond culture—space, identity, and the politics of difference. *Cultural Anthropology, 7*(1), 6–23.

Haalboom, B., & Natcher, D. C. (2013). The power and peril of "vulnerability": Lending a cautious eye to community labels. In D. C. Natcher, R. C. Walker, & T. S. Jojola (Eds.), *Reclaiming indigenous planning* (Vol. 357, pp. 357–375). Montreal: McGill-Queen's University Press.

Hann, C. M. (2002). All Kulturvölker now? Social Anthropological Reflections on the German-American Tradition. In R. G. Fox and B. J. King (Eds.), *Anthropology Beyond Culture* (pp. 221–238). Oxford: Berg

Hawkes, J. (2001). *The fourth pillar of sustainability. Culture's essential role in public planning*. Victoria: Cultural Development Network.

Hovelsrud, G. K., White, J. L., Andrachuk, M., & Smit, B. (2010). Community adaptation and vulnerability integrated. In G. K. Hovelsrud & B. Smit (Eds.), *Community adaptation and vulnerability in arctic regions* (pp. 335–348). Dordrecht: Springer.

Ignatieva, S. S., Vinokurova, U. A., Mestnikova, A. E., Vinokurova, E. P., Zaharova, A. E., Chusovskaia, V. A., . . Nabok, I. L. (2014). *Kul'tura Arktiki. Kollektivnaya monografiia/pod obshhey redakciey*. Yakutsk: Seriia Kul'tura Arktiki.

Ingold, T. (2018). *Anthropology: Why it Matters*. Cambridge: Polity Press.

Jacob, M. M. (2013). *Yakama rising. Indigenous cultural revitalization, activism, and healing, first peoples: New directions in indigenous studies*. Tucson: The University of Arizona Press.

Kalutskov, V. N. (2008). *Landshaft v kul'turnoi geografii*. Moscow: Novyi khronograf.

Kirsch, S. (2001). Lost worlds, environmental disaster, 'culture loss,' and the law. *Current Anthropology, 42*(2), 167–198.

Kivitalo, M., Kumpulainen, K., Soini, K., Dessein, J., Battaglini, E., & Horlings, L. (2016). Exploring culture and sustainability in rural Finland. In *Cultural sustainability and regional development: Theories and practices of territorialisation* (pp. 94–107). London: Routledge.

Klokov, K. B. (1998). *Traditsionnoe prirodopol'zovanie korennyh malochislennyh narodov Severa (geograficheskie i social'no-ekonomicheskie problemy). Dissertaciia na soiskanie stepeni doktora geograficheskih nauk*. Moskva: Institut geografii RAN.

Klokov, K. B. (2011). Traditsionnoe severnoe olenevodstvo v kontekste ustoichivogo razvitiia Severa Rossii. *Izvestiia Sankt-Peterburgskogo gosudarstvennogo agrarnogo universiteta, 25*, 131–135.

Klokov, K. B. (2012). *Sovremennoe polozhenie olenevodov i olenevodstva v Rossii*. Sever i severiane. (2012). Sovremennoe polozhenie korennyh malochislennyh narodov Severa, Sibiri i Dal'nego Vostoka Rossii / ed. Novikova N. I. & Funk D. A. Moskva: IEA RAN. 36–38.

Klokov, K. B., Krasovskaya, T. M., & Krasovskaya, A. N. (2001). *Problemy perekhoda k ustoichivomu razvitiiu rajonov rasseleniia korennyh narodov rossiiskoi Arktiki*. Moscow: RAN.

Kohn, E. (2015). Anthropology of ontologies. *Annual Review of Anthropology, 44*, 311–327.

Komarova, N. G. (2014). *Industrial'noe nasledie v landshafte i kul'ture rossiiskogo severa (geoekologicheskii analiz) V sbornike: Sergeevskie chteniia Yubileinaia konferentsiia, posvyashhennaya 100-letiiu so dnya rozhdeniia akademika E.M. Sergeeva* (pp. 282–284). Materialy godichnoi sessii Nauchny sovet RAN po problemam geoekologii, inzhenernoi geologii i gidrogeologii: IGJe RAN.

Kriukov, V. A., Shishackii, N. G., Briuhanova, E. A., Kobalinskii, M. V., Matveev, A. M., & Tokarev, A. N. (2014). *Potencial ustoichivogo razvitiia arealov prozhivaniia i ekonomicheskaya ocenka kachestva zhizni korennyh malochislennyh narodov Severa.* Novosibirsk: Institut ekonomiki i organizacii promyshlennogo proizvodstva SO RAN.

Lamazhaa, C. K. (2010). Krugly stol "Sociokul'turny neotraditsionalizm i ustoichivoe razvitie regionov Rossii". *Novye issledovaniia Tuvy, 2,* 14–24.

LaRocque, E. (2010). *When the Other is Me: Native resistance discourse, 1850–1990.* Winnepeg: University of Manitoba Press.

Loach, K., Rowley, J., & Griffiths, J. (2016). Cultural sustainability as a strategy for the survival of museums and libraries. *International Journal of Cultural Policy, 1–16,* 186–198.

Lopinto, N. (2009, June). Der Indianer: Why do 40,000 Germans spend their weekends dressed as native Americans? In *UTNE reader.* Retrieved from www.utne.com/mind-and-body/Germans-weekends-Native-Americans-Indian-Culture.

Low, S. M. (2008). Social sustainability. In G. Fairclough, R. Harrison, H. J. Jameson, & J. Schofield (Eds.), *The heritage reader* (pp. 392–404). New York, NY: Routledge.

Lukin, J. F. (2011). Perfomans etnokul'turnogo landshafta Arktiki v global'nom i regional'nom izmereniiah. *Arktika i Sever, 1,* 56–88.

Lukin Ju.F. (2015a). "Etnokul'turny landshaft rossiyskoi Arktiki: ot konceptualizacii znanii k upravleniiu konfliktami." *Arktika i Sever, 21,* 118–143.

Lukin Ju.F. (2015b). "Obespechenie bezopasnosti i ustoichivogo razvitiia arkticheskogo regiona, sohranenie ekosistem i traditsionnogo obraza zhizni korennogo naseleniia Arktiki." *Arktika i Sever, 21,* 190–197.

Mangataeva, D. (2000). *Evoliutsiia traditsionnykh sistem zhizneobespecheniia korennykh narodov Baikal'skogo regiona.* Novosibirsk: Izdatel'stvo Sibirskogo otdeleniia RAN.

Marfusalova, A. D. (2002). *Mudrost' ekotraditsii severyan.* Yakutsk: Izdatelstvo SO RAN.

Meitz, A., & Gartler, S. (2017). Arctic FROST young scholars panel—Arctic anthropology and sustainability. *The Polar Journal, 7*(1), 249–250.

Meitz, A., & Ringhofer, K. (2017). The bicycle and the Arctic—resilient and sustainable transport in times of climate change. In *The interconnected Arctic—UArctic congress* (pp. 157–164). New York: Springer.

Mikhailova, G. V. (2014). Ostrov Kolguev: prirodnoe i kul'turnoe nasledie kak faktor ustoichivogo razvitiia korennyh narodov Arktiki. *Sovremennye problemy nauki i obrazovaniia, 6,* 1571.

Myllyviita, T., Lähtinen, K., Hujala, T., Leskinen, L., Sikanen, L., & Leskinen, P. (2014). Identifying and rating cultural sustainability indicators: A case study of wood-based bioenergy systems in eastern Finland. *Environmental Development and Sustainability, 16*(2), 287–304.

Nadasdy, P. A. (2008). *Wildlife as renewable resource. Timely assets. The politics of resources and their temporalities* (E. E. Ferry & M. E. Limbert, Eds.). Santa Fe: SAR Press.

Nagel, J. (1996). *American Indian ethnic renewal: Red power and the resurgence of identity and culture.* New York, NY: Oxford University Press.

Natcher, D. C., & Davis, S. (2007). Rethinking devolution: Challenges for aboriginal resource management in the Yukon Territory. *Society and Natural Resources, 20*(3), 271–279.

Natcher, D. C., & Haalboom, B. (2013). The power and peril of "vulnerability": Lending a cautious eye to community labels. In D. C. Natcher, R. C. Walker, & T. S. Jojola (Eds.), *Reclaiming indigenous planning* (Vol. 357, p. 375). Montreal: McGill-Queen's University Press.

Nations, U. (2019). *Sustainable development goals*. Retrieved from www.un.org/sustainabledevelopment/sustainable-development-goals/

Neufeld, D. (2016). Our land is our voice: First nation heritage-making in the Tr'ondek/Klondike. *International Journal of Heritage Studies, 22*(7), 568–581.

Overland, I. (2009). Indigenous rights in the Russian North. In E. W. Rowe (Ed.), *Russia and the north* (pp. 165–186). Ottawa: University of Ottawa Press.

Peylasov, A. N. (2009). *And the last becomes the first: Russian periphery on the way to knowledge economy.* Moscow: Librokom.

Peter, D. L., & Hogan, J. (2006). First nation of Na-Cho Nyäk Dun lands and resources department. In L. R. Bleiler, C. Burn, & M. O'Donoghue (Eds.), *Heart of the Yukon—A natural and cultural history of the Mayo area* (pp. 82–133). Mayo: Village of Mayo.

Petrov, A. N., BurnSilver, S., Chapin, S. F., Fondahl, G., Graybill, J., Keil, K., . . Schweitzer, P. (2016). Arctic sustainability research: Toward a new agenda. *Polar Geography, 39*(3), 165–178.

Petrov, A. N., BurnSilver, S., Chapin, S. F. III, Fondahl, G., Graybill, J., Keil, K., . . Schweitzer, P. (2017). *Arctic sustainability*. London: Research, Routledge Focus.

Pika, A. (1996). *Anxious North: Indigenous peoples in Soviet and post-Soviet Russia.* Copenhagen: IWGIA.

Pisareva, T. M., & Vinogradov, A. N. (2010). *Muzey kak tsentr vospitaniia kul'tury prirodopol'zovaniia na Krajnem Severe. Trudy Fersmanovskoi nauchnoi sessii, 7*, 196–199. GI KNC RAN.

Pivneva, E. A. (2015). *Dinamika traditsii v arkticheskom izmerenii, 47*(2), 98–107.

Polistina, K. (2009). Cultural literacy: Understanding and respect for the cultural aspects of sustainability. In A. Stibbe (Ed.), *The handbook for sustainability literacy. Skills for a changing world.* Oxford: Green Books.

Radkau, J. (2011). *Die Ära der Ökologie: Eine Weltgeschichte.* München: Beck.

Ragulina, M. V. (2000). *Korennye etnosy sibirskoi tajgi: motivaciia i struktura prirodopol'zovaniia (na primere tofalarov i evenkov Irkutskoi oblasti).* Novosibirsk: Izdatelstvo SO RAN.

Riabova, L. (2010). *Community viability and wellbeing in the circumpolar north.* Fairbanks: University of Alaska Press.

Robbek, V. A. (2000). *Tsirkumpoliarnaia Kul'tura—nevostrebovanny rezerv chelovecheskoi civilizacii. Cirkumpoliarnaya kul'tura: pamyatniki kul'tury narodov Arktiki i Severa* (pp. 3–8). Yakutsk: Severoved.

Robbek, V. A. (2010). Traditsionnye znaniia narodov Arktiki i Severa—rezerv usto-ichivogo razvitiia Arktiki. *Voprosy istorii i kul'tury severnyh stran i territorii, 4*(12), S. 10–13.

Rodoman, B. (2015). Sokhranim russkuiu derevniu! (khotia by v kachestve muzeia-zapovednika). In *Puti Rossii. Al'ternativy.* (pp. 389–396). Moskva: Novoe litera-turnoe obozrenie.

Romanova, E. N., & Nikiforova, V. S. (2015). *Kul'turny landshaft traditsionnykh obshhestv v kontekste riskogennogo prostranstva rossiiskoi Arktiki (resursy ustoichivogo razvitiia). Sotsial'nye riski i upravleniie imi v sovremennom obshchestve.* Proceedings of the All-Russian scientific-applied conference. December 5, 2014. Ed. by E. N. Shovin., 36–38.

Russian Federation. (2001). *O territoriiakh traditsionnogo prirodopol'zovaniia korennykh malochislennykh narodov Severa, Sibiri i Dal'nego Vostoka Rossiiskoi Federatsii* (No. Federal'nyi zakon ot 7 maia 2001 g. N 49-FZ). Retrieved from Kremlin website: http://base.garant.ru/12122856/

Sahlins, M. (1999). Two or three things that I know about culture. *Journal of the Royal Anthropological Institute, 5,* 399–421.

Schweitzer, P., Fox, S. I., Csonka, Y., & Kaplan, L. (2010). *Cultural wellbeing and cultural vitality. In Arctic social indicators—a follow-up to the Arctic human development report* (J. N. Larsen, P. Schweitzer, & G. Fondahl, Eds.). Copenhagen: Nordic Council of Ministers.

Schweitzer, P., & Ulturgasheva, O. (2014). Cultures and identities. In J. N. Larsen (Ed.), *Arctic human development report.* (pp. 104–145). Copenhagen: Nordisk Ministerråd.

Shapkhaev, S. G., & Shapkhaev, B. S. (2005). Traditsionnoe prirodopol'zovanie korennyh narodov mira kak faktor ustoichivogo razvitiia. Korennye narody Buriatii v nachale XXI veka. Materialy regional'noi nauchno-prakticheskoi konferencii. *A.A* (pp. 50–94). Ulan-Ude: Izd. BGSHA im. V.R. Filippova.

Sirina, A. N. (2008). Chuvstvuiushhie zemliu: ekologicheskaia etika evenkov i evenov. *Etnograficheskoe obozrenie, 2,* 121–138.

Snow, K., & Ochlaski, H. (2018). Making room: Cultural resistance through Pilimmaksarniq. *Education in the North, 25,* 3–32.

Soini, K., & Birkeland, I. (2014). Exploring the scientific discourse on cultural sustainability. *Geoforum, 51,* 213–223.

Soini, K., & Dessein, J. (2016). Series introduction, cultural sustainability, and regional development: Theories and practices of territorialization. In *Cultural sustainability and regional development* (pp. xiii–xiv). London: Routledge.

Stammler-Gossmann, A. (2010). Translating' vulnerability at the community level: Case study from the Russian North. In B. Smit & G. K. Hovelsrud (Eds.), *Community adaptation and vulnerability in Arctic regions* (pp. 131–162). Berlin: Springer.

Stolcke, V. (1995). Talking culture: New boundaries, new rhetoric's of exclusion in Europe. *Current Anthropology, 36*(1), 1–24.

Stylianou-Lambert, T., Boukas, N., & Christodoulou-Yerali, M. (2014). Museums and cultural sustainability: Stakeholders, forces, and cultural policies. *International Journal of Cultural Policy, 20,* 566–587. https://doi.org/doi:10.1080/10286632.2013.874420

Swyngedouw, E. (2010). Conceptual challenges for planning theory. In J. Jillier & P. Healey (Eds.), *Conceptual challenges for planning theory* (pp. 299–320). Farnham: Ashgate.

Throsby, D. (1995). Culture, economics and sustainability. *Journal of Cultural Economics, 19*(3), 199–206.

Tiugashev, E. A. (2010). Ekonomika ustoichivogo razvitiia: vozrozhdenie tsennostey traditsionnoi ekonomiki. *Novye issledovaniia Tuvy, 2*, 50–61.

Tomaska, A. G. (2017). Migranty Respubliki Saha (Yakutiia): voprosy integracii detey sel'skih migrantov. *Severo-Vostochny gumanitarny vestnik, 1*(18), 58–64.

Trouillot, M. R. (2002). Adieu, culture: A new duty arises. In R. G. Fox & B. J. King (Eds.), *Anthropology beyond culture*. Oxford: Berg.

Truievtseva, O. N. (2012). Muzei Sibiri i sohranenie kul'turnogo naslediia korennyh narodov. *Vestnik Kemerovskogo gosudarstvennogo universiteta kul'tury i iskusstv, 19*(2), 76–82.

Turaev, A. E. (1997). *Territoriia traditsionnogo prirodopol'zovaniia: pravovye osnovy, status, opyt organizatsii. Etnos i prirodnaia sreda.* Vladivostok: Dal'nauka.

UNESCO. (2016). *Culture: Urban future. Global report on culture for sustainable urban development.* Retrieved from http://unesdoc.unesco.org/images/0024/002459/245999e.pdf

Urquhart, D. (2013). *Nena Doòli. Fish and Wildlife Dooli and Traditional Laws, Little Salmon/Carmacks First Nation and Selkirk First Nation.*

Vedenin, J. A. (2003). Informacionnye osnovy izucheniia i formirovaniia kul'turnogo landshafta kak ob'ekta naslediia. *Izvestiia Akademii Nauk: Seriia Geograficheskaya, 3*, 7–13.

Vernadsky, V. I. (1991). *Nauchnaya mysl' kak planetnoe iavlenie.* Moscow: Nauka.

Vinokurova, L. I. (2003). Zhenshhina Severa v epokhu peremen. In *Sever: obshhestvo, etnosy, chelovek. SO RAN* (pp. 92–99). Yakutsk: YaF Izdatelstva SO RAN.

Vinokurova, U. A. (1997). *Vospitanie i obrazovanie detei narodov Severa.* Yakutsk: Ministerstvo obrazovaniia.

Vinokurova, U. A., & Yakovets, U. V. (2016). *Arctic circumpolar civilization. Educational edition.* Novosibirsk: Nauka.

Vinokurova, U. A., Zhegusov, U., Mestnikova, A. E., Alekseeva, G. G., & Alekseev, V. N. (2017). Mezhdunarodnye sportivnye igry "Deti Azii" kak sotsiokul'turnyi proekt Respubliki Sakha (Yakutia). *Teoria i praktika fizicheskoi kul'tury, 3*, 94–96.

Yencken, D., & Wilkinson, D. (2001). *Resetting the compass: Australia's journey towards sustainability.* Collingwood: CSIRO Publishing.

4　Sustainable resources

Chris Southcott

The difficult question of sustainable resources in the Arctic

An early attempt to apply the concept of sustainable development to the Arctic was by Duerden (1992), who recognized its desirability as "an easily articulated basis for empowerment and action for communities and populations that have felt the adverse environmental and social impacts of resource exploitation" (p. 219). Duerden noted that early research stressed that Indigenous communities were inclined toward sustainability, which fit well with their notions of development. In early publications, 'sustainable development' is synonymous with increased local community empowerment, more environmentally sound practices, and "local investment of non-renewable resource revenues" (p. 219).

Duerden also explained the socio-historic conditions that made sustainable development essential for northern communities and their role in this type of development. He also outlines concerns with some approaches to sustainable development in the Canadian Arctic, including the importance of land-based subsistence economies and the possibility that these activities may cease. For example, he notes that traditional harvesting activities may include unsustainable food practices in which greater calories are consumed than returned. Because Indigenous communities rely on non-renewable resources to survive, they are innately unsustainable. Thus, northern sustainable development must be seen from a relative and not an absolute perspective as well as from a local scale. Regarding extractive industries in the North, such activities should continue for the survival of northern communities. It follows logically that sustainable northern non-renewable resource development must ensure that more benefits from extractive activities stay in the region. They must be subsidized and support subsistence activities.

Non-renewable resources and sustainability

Mining and oil and gas development comprise the two main Arctic extractive industries. Mining is more closely related to the sustainability concept than hydrocarbons, but questions remain about whether mining, based on extraction of a limited mineral, can become sustainable (Emberson-Bain, 1994). Some consider that mining, while not sustainable, could become more sustainable or could provide resources to assist sustainability (Beckerman, 1994). Realizing these visions mean that, first, mining activities must ensure minimal environmental damage, possible with advanced mining techniques. Second, lasting economic benefits for future generations must be ensured, especially for communities directly impacted. When discussions about mining and sustainability emerged, economists noted that increased local benefits tended to negative affect other economic sectors, called the Dutch disease (Auty & Warhurst, 1993). This perspective developed into the resource curse discourse (Auty, 1994; Sachs & Warner, 2001). The biggest problem for mining in becoming sustainable was devising a macroeconomic policy to address the resource curse and related problems (Stiglitz, 2007).

Scholars also note problems defining and measuring sustainable development in peripheral regions dependent on extractive industry (Davis, 1996; Cragg, 1998), and some called for a sustainable development based on justice and the needs of local populations (Low & Gleeson, 1998). By the late 1990s, the mining industry began to address sustainable development more seriously (Bhattacharya, 2000; Hilson & Murck, 2000; Humphreys, 2001). The Whitehorse Mining Initiative (Fitzpatrick, Fonseca, & McAllister, 2011) reported on sustainable mining and led to the Sustainable Development Framework in the international mining industry.

Criticism of the mining industry continues today and includes debates about indicators for monitoring sustainable mining (Aguado & Nicieza, 2008; Horsley, Prout, Tonts, & Ali, 2015) and how to strengthen communities. Companies had to ensure communities, especially Indigenous, that they were sustainable (Lertzman & Vredenburg, 2005; Richards, 2009). This stimulated interest in socioeconomic sustainability indicators (Nelsen, Scoble, & Ostry, 2010; Nygaard, 2016), including gender (Lahiri-Dutt, 2011). The idea of a "social license to operate" conceptualized greater community benefits and a greater say in decision making (Prno & Slocombe, 2012; Moffat, Lacey, Zhang, & Leipold, 2016).

Discourse on sustainability and mining highlights several themes, one of which points to internal industry differences. Large-producing companies generally supported sustainability, but others—especially smaller exploration companies—were less concerned (Thomson & Joyce, 2006). Discussion ensued about supply chains and how to use them to provide community benefits (Lydall, 2009) making supply chain sustainability important to mining sustainability (Fleury & Davies, 2012) as well as peak minerals concept (Prior, Giurco, Mudd, Mason, & Behrisch, 2012). Industry sustainability reporting, including corporate social responsibility, is important (Dashwood, 2014; Boiral, 2013; de Villiers, Low, & Samkin, 2014; Fonseca, McAllister, & Fitzpatrick, 2014). Starting in 2013, critical attention to sustainability reporting became a now-growing field of corporate social responsibility.

The hydrocarbon industry seems less concerned with sustainability (Escobar & Vredenburg, 2011). General discussions note that this development is unsustainable because of finite reserves or environmental–climate impacts (Bradshaw, 2010; MacKenzie, 1997). Discourse about sustainable development occurs with such concepts as "supply chain sustainability" and networks (Hall, Matos, & Silvestre, 2012). Some scholarship assesses how much hydrocarbons contribute to sustainable development (Bebbington, Brown, & Frame, 2007; Ediger, Hosgor, Surmeli, & Tatlidil, 2007).

Renewable resources and sustainability

Three main Arctic renewable resource activities are addressed: commercial fishing, commercial reindeer herding, and traditional subsistence harvesting. Although not always practiced, sustainability is central to commercial fisheries (Bailey, 1996; Drummond & Symes, 1996). Literature from the 1990s, noted the unsustainability of industrial fishing (Clapp, 1998; Otterstad, 1996). Discussions highlighted fisheries management (Caddy, 1999; Chiou, Tzeng, & Cheng, 2005), using the notion of sustainable yield (Glen, 1995; Hilborn, 2010). Recent debate about the long-term fisheries sustainability addresses concepts of community versus individual quotas (Symes & Crean, 1995; Wingard, 2000), the sustainable livelihoods approach (Allison & Horemans, 2006), ecosystem management (Gilman, Passfield, & Nakamura, 2014), and aquaculture (Costa-Pierce, 2010). Fisheries sustainability discourse notes the importance of fishing for Indigenous communities, creating research that discusses traditional subsistence activities (Adams, 1993; Akimichi, 1995).

Review of Arctic literature

Against this background, a bibliometric analysis was undertaken using the Web of Science database to examine sustainability discourse about Arctic natural resources systematically. Searches were limited to social sciences and humanities. Customized settings included only the Social Science Citation Index (1975–present), Arts and Humanities Citation Index (1975–present), Conference Proceedings Citation Index—Social Science and the Humanities (1990 present), and Emerging Sources Citation Index (2015–present) (Table 4.1). Based on a review of titles, keywords, and abstract, an initial list of 179 articles was established, which was reduced to a final 152 publications. Final searches concluded on July 6, 2018 (Figure 4.1).

Table 4.1 Words included in bibliometric searches

Theme	Activity	Geography
Sustainable, sustainability	Natural resources, mining, oil and gas, fisheries, resource development	Arctic, North, Circumpolar North, Alaska, Yukon, Northwest Territories, Nunavut, Nunavik, Nunatsiavut, Labrador, Greenland, Iceland, Faroes, Norway, Finnmark, Sweden, Norrbotten, Finland, Lapland, Russia, Siberia, Kola, Arkhangelsk, Yamal, Nenets, Krasnoyarsk, Sakha, Chukotka, Magadan, Kamchatka, Sakhalin

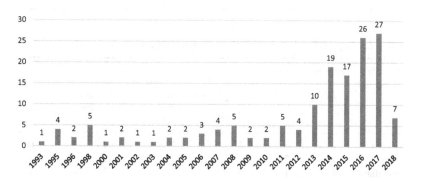

Figure 4.1 Chart of publications about sustainability by year. The number of publications started to increase significantly in 2013 with only 30% of works being published between 1993 and 2012. The year that produced the largest number of publications dealing with sustainability and resources in the Arctic was 2017, but it is important to note that numbers for 2018 are only partial.

Changes in meanings of sustainability

Analysis indicates varying meanings of sustainability and sustainable development seem to be based on the type of resource discussed. For example, subsistence harvesting was not a primary area of analysis, but articles about it provide understanding of sustainability concepts varied based on the natural resources investigated, enabling understanding of some aspects of sustainability and natural resource development (Tables 4.2 and 4.3).

Table 4.2 Main journals included in the bibliometric analysis

Journal name	Number of publications	Percent of all publications
Marine Policy	19	12.5
Ecology and Society	9	5.9
Society & Natural Resources	7	4.6
Ekonomika Regiona— Economy of Region	6	3.9
Arctic	5	3.3
Sustainability	5	3.3
Extractive Industries and Society: An International Journal	4	2.6
Human Ecology	4	2.6
Coastal Management	3	2.0
Polar Record	3	2.0
Resources Policy	3	2.0

Table 4.3 Regional and resource activity foci of publications

Activities by region

	Fishing	Mining	Oil and gas	Traditional subsistence activities	Resource development	Total
Alaska	27	3	3	7	4	44
Canada	3	2	2	2	5	14
Greenland	1	1	0	2	2	6
Iceland	9	0	0	0	0	9
Norway	12	1	1	1	2	17
Sweden	0	1	1	1	2	5
Finland	0	5	0	0	1	6
Russia	5	3	10	3	18	39
Arctic	2	2	2	0	6	12
Total	59	18	19	16	40	152

Importance of local culture for sustainability

An early article discussing Arctic sustainability examined the sustainability of Indigenous subsistence whaling in Greenland (Caulfield, 1993). Despite criticisms of the sustainability of Indigenous whaling, Caulfield claimed its importance for long-term community sustainability. In addition to ensuring that communities do not become overly dependent on the external world economy, these activities sustain cultural traditions. Others find sustainability linked to maintaining Indigenous cultures (Davis & Jentoft, 2001), and using local, traditional knowledge retains information essential to resource management (Armitage, 2005). These core ideas persist, stressing the importance of traditional harvesting for food security and cultural maintenance and promoting traditional knowledge's values (Gerlach & Loring, 2013; Reedy-Maschner & Maschner, 2013; Thornton, 2015). Sustainability of reindeer herding is also an important topic (Forbes, 2013; Klokov & Khrushchev, 2016).

Non-renewable impacts on the environment

Initial research on sustainability and extractive resources in the Arctic stressed their incompatibility. Early works highlighted that sustainability was threatened by extractive development via environmental destruction, making life difficult for Indigenous peoples (Bradshaw, 1995; Chance & Andreeva, 1995). Often such articles addressed post-Soviet Russia (Glazyrina, 1998). Another perspective highlights how existing political and economic structures and their relationship to natural resource development produced poverty and a lack of economic development in Arctic communities (Berardi, 1998).

Extractives as a resource for Arctic sustainability

Research emerged in the late 1990s that examined how management of extraction allowed Arctic communities to become sustainable (Eglington, Israel, & Vartanov, 1998; Glazyrina, 1998). This tendency continues in Russia (Baklanov & Moshkov, 2016; Korobitsyn, 2015), Greenland (Tiainen, 2016), and North America (Huskey & Southcott, 2016; Rodon, Levesque, & Blais, 2013). This research indicates barriers to sustainability (van Bets, van Tatenhove, & Mol, 2016), in which common themes are the importance of local empowerment, local involvement, and local knowledge (Wilson & Stammler, 2016).

Unique to Arctic Russia, researchers note the tendency for company responsibility for environmental management (Ray, 2008; Salmi, 2008;

Shvarts, Pakhalov, & Knizhnikov, 2016). How companies negotiate social licenses to operate to enhance sustainability concerns researchers in Russia, in North America, and Fennoscandia (Prno, 2013; Riabova & Didyk, 2015). Related is the ability of local populations to become involved in formal impact assessment processes (Nygaard, 2016; Pettersen & Song, 2017), how financial capital can be managed for long-term sustainability (Baena, Sevi, & Warrack, 2012; Nikolakis, Nelson, & Cohen, 2014), and the resource curse (Suutarinen, 2015; Tynkkynen, 2007; Wennecke, 2017). Many studies indicate that sustainability is dependent on dealing with conflicts associated with extractive resource development (Lindahl, Johansson, Zachrisson, & Viklund, 2018; Sairinen, Tiainen, & Mononen, 2017), and benefit sharing has emerged as a focus (Tulaeva & Tysiachniouk, 2017; Tysiachniouk & Petrov, 2018; Wilson & Stammler, 2016).

Sustainability and the fishing industry

Much literature about Arctic sustainability and natural resources addresses fishing. Here, sustainability is linked to the continued existence of available fish stocks, commercially and otherwise. Discussions highlighted how fish stocks could be exploited without threatening their continued existence (Gullestad et al., 2017) in which the maximum possible level of exploitation is called sustainable yield (Hilborn, Parrish, & Litle, 2005). Many articles addressed best or sustainable management of fish stocks through policies such as individual transferable quotas (Eythorsson, 1996; Symes & Crean, 1995). Sustainability of a particular stock is linked to the long-term sustainable development of actual human communities (Anderson et al., 2017; Chambers & Kokorsch, 2017).

Some research examined industry from the perspective of sustainable community development. Ween and Colombi (2013) find two distinct conceptualizations of sustainability: cultural and biological. Others note that a patriarchal fishing industry threatened sustainability (Munk-Madsen, 1998), or the conflict between commercial and "local" fisheries and how to manage these conflicts for sustainability or community resilience (Kyllonen et al., 2006; Robards & Greenberg, 2007), or suggesting that local fisheries were more likely to be sustainable with threats such as climate change (McGoodwin, 2007).

A common theme in the fisheries literature is the difficulty in balancing local versus non-local commercial fisheries (Carothers, 2008; Langdon, 2008). Interest in examining the impacts of varying types of certification of sustainable fisheries (Foley & Hebert, 2013; Gulbrandsen & Honneland, 2014) is clear, with many scholars focusing on the Marine Stewardship Council.

Crosscutting themes

While definitions and themes of sustainability differ according resource type, several crosscutting themes exist. Almost universally accepted is the idea that local control must be enhanced for sustainability to occur (Chapin, Sommerkorn, Robards, & Hillmer-Pegram, 2015; Gerlach & Loring, 2013). Adequate local involvement is stressed by some (Riabova & Didyk, 2015; Stetson & Mumme, 2016), whereas others stress the integration of local traditional knowledge into these processes (Armitage, 2005; Parlee, Sandlos, & Natcher, 2018).

The impacts of climate change on sustainable resource development is another crosscutting theme (Arruda & Krutkowski, 2017; Forbes & Stammler, 2009; Harsem & Hoel, 2013). This research concludes that climate change makes resource development more impactful on Arctic communities, thereby making community empowerment fundamental for creating long-term regional sustainable development.

Less prevalent in the 2000s is the theme of conflict between extractive industry and sustainable development. Fewer researchers discuss these two concepts as oppositional, but research examining difficulties between Indigenous subsistence activities and extractive resource development continues (Gunn, Russell, & Greig, 2014; Parlee et al., 2018; Southcott & Natcher, 2018). The need for improved methodological frameworks and indicators is noted (Davankov, Dvinin, & Postnikov, 2016; Johannesson, Daviosdottir, & Heinonen, 2018; Korobitsyn, 2015), especially the increased use of qualitative data and methodologies attuned to Indigenous communities (Lyons, Carothers, & Reedy, 2016).

Theoretical and conceptual perspectives

Theoretical and conceptual perspectives about sustainability and resources in the Arctic do not seem to be related to the types of resources analyzed. Two separate theoretical and conceptual discourses exist. The first, the political economy perspective, is inspired by social scientists. The second, the social ecological systems perspective, integrates concepts derived partially from the natural sciences.

First, the political economy perspective includes multiple conceptual frameworks that tend to include economic and political contexts for sustainability (Baklanov & Moshkov, 2016; Hebert, 2015; Novoselov, Potravny, Novoselova, & Gassiy, 2017). These works use standard economic conceptualizations but address Arctic resources, meaning they integrate ecological or environmental factors into their analyses (Davankov et al., 2016; Dudin, Lyasnikov, Protsenko, & Tsvetkov, 2017). Additionally, the "capital approach" (Uwasu & Yabar, 2011), derived from economic theory, evaluates the balancing of human, financial, social, and natural capital (Belonozhko, Silin, & Gyurdzhinyan, 2018; Korobitsyn, 2015).

Political economy frameworks agree that environmental factors are also important to address. Recent publications increasingly use political ecology (Loring, 2017; Thornton & Hebert, 2015) to describe existing power structures, dispossession of land and resources from Indigenous peoples, and implications for sustainable development (Gallardo et al., 2017).

Second, a social-ecological systems (SESs) perspective provides more coherence, borrowing systems-related concepts from natural sciences to integrate social and environmental concerns (Fauchald, Hausner, Schmidt, & Clark, 2017). Use of the framework varies: some highlight ecosystem management (Dahl, Ervik, Iversen, Moksness, & Saetre, 2008; Mangel, 2010), yet others examine ecosystem stewardship (Chapin et al., 2015; Hansen, 2014), risk management (Blair, Lovecraft, & Kofinas, 2014), or resilience (Chapin et al., 2015; Forbes, 2013; Himes-Cornell & Hoelting, 2015). From the SES approach, a sustainability science emerges for studying resource-based sustainability (Ray, Kolden, & Chapin, 2012).

State of sustainability

In summarizing the state of sustainability for Arctic resources, it is best to examine each by the type of resources. Again, while subsistence harvesting was not a central focus of this review, much of the literature dealing with other resources discussed the sustainability of these activities. Despite Lovins' comments in Duerden (1992), most research stresses that subsistence harvesting remains essential to Arctic community sustainability (Pelyasov, Galtseva, & Atamanova, 2017). Some suggest that traditional activities may be endangered by other types of resource development (Parlee et al., 2018), but there is general agreement that such activities contribute to long-term Arctic community sustainability by enhancing food security and cultural sustainability (Gerlach & Loring, 2013; Ween & Colombi, 2013).

The discourse surrounding sustainability and fishing is more advanced than for other types of resources. This discourse is often centered on the sustainability of a particular species for harvesting rather than the sustainability of human communities dependent on these species for their long-term sustainable development. Some researchers have begun examining the impacts of various fisheries management policies on Arctic fishing community sustainability (Chambers & Kokorsch, 2017; Ksenofontov, Backhaus, & Schaepman-Strub, 2017), but this work is limited to issues such as certification schemes (Foley & Havice, 2016).

For extractive resource activities such as mining and oil and gas development, the literature has moved from a general idea of incompatibility between these types of developments and the sustainable community development (Berardi, 1998; Chance & Andreeva, 1995) to a reexamination of the potential for extractive activities to assist the sustainability of these communities where other opportunities are lacking (Rodon & Schott, 2014). Research now more critically examines potential benefits arising from mining and oil and gas developments to examine how they can help achieve sustainable futures (Southcott, Abele, Natcher, & Parlee, 2018).

Literature on best practices for sustainability is emerging. Economically, comparisons of financial benefits from extractive industry without causing problems associated with the resource curse by using sovereign wealth funds. The Norwegian Government Pension Fund and the Alaska Permanent Fund are cited as a successful example (Baena et al., 2012; Ramirez-Cendrero & Wirth, 2016). For extractive resource development, a successful mine for promoting regional sustainability is Alaska's Red Dog Mine because local communities have participated in mine planning, ensuring that they receive real benefits (Kadenic, 2015; Prno, 2013). Regarding fisheries, Alaska's community quota system indicates effective ways to ensure long-term community sustainability. General regulatory systems designed around individual quotas remains problematic (Langdon, 2008), but Indigenous communities needs are expressed (Lyons et al., 2016).

Knowledge gaps and future priorities

Gaps and priorities should be based on the needs of Arctic communities. From this perspective, what is required for future research? First, it is necessary to understand the barriers to subsistence activities (Natcher, Shirley, Rodon, & Southcott, 2016). Limited empirical research examines assumptions about the limits to subsistence practices. Industrial development advancement, which appears to negatively impacts subsistence, is complex because certain conditions benefit it (Southcott & Natcher, 2018). Future research must concentrate on finding the best way to support subsistence activities, including using potential benefits provided by extractive industry developments. Second, understanding the impact of fishing on community sustainability (Robards & Greenberg, 2007) is necessary. Knowledge of fish stocks and management policies, examination of policy impacts on the local communities, and the use of community fishing quotas are promising areas of study. Third, examination of extractive industries' regulatory systems, attention to local empowerment through self-government, and worldwide recognition of Indigenous rights means greater possibility of increasing regional sustainability (Tiainen, 2016). More research is needed about negative and cumulative impacts of extraction and how to mitigate them (Gunn et al., 2014).

Three crosscutting priorities emerge. First, consensus that traditional knowledge is important for the Arctic cultural sustainability exists, and the traditional knowledge should be used in planning, operating, and closing resource developments (Huntington, 2014). Second, social innovation should assist sustainable Arctic resource development (Kryukov, Sevastyanova, Tokarev, & Shmat, 2017; Wennecke, 2017). Finally, there is a need for informed and effective research methodologies informed by Arctic communities (Lyons et al., 2016).

References cited

Adams, W. M. (1993). Indigenous use of wetlands and sustainable development in West Africa. *Geographical Journal, 159*, 209–218.

Aguado, M. B. D., & Nicieza, C. G. (2008). An empirical index to evaluate the sustainability of mining projects. *International Journal of Environment and Pollution, 33*(2–3), 336–359.

Akimichi, T. (1995). Indigenous resource management and sustainable development: Case studies from Papua New Guinea and Indonesia. *Anthropological Science, 103*(4), 321–327.

Allison, E. H., & Horemans, B. (2006). Putting the principles of the sustainable livelihoods approach into fisheries development policy and practice. *Marine Policy, 30*(6), 757–766.

Anderson, S. C., Ward, E. J., Shelton, A. O., Adkison, M. D., Beaudreau, A. H., Brenner, R. E., & Williams, B. C. (2017). Benefits and risks of diversification for individual fishers. *Proceedings of the National Academy of Sciences of the United States of America, 114*, 10797–10802.

Armitage, D. R. (2005). Community-based Narwhal management in Nunavut, Canada: Change, uncertainty, and adaptation. *Society & Natural Resources, 18*(8), 715–731.

Arruda, G. M., & Krutkowski, S. (2017). Social impacts of climate change and resource development in the Arctic implications for Arctic governance. *Journal of Enterprising Communities-People and Places of Global Economy, 11*(2), 277–288.

Auty, R. M. (1994). The resources curse thesis: Minerals in Bolivian development, 1970–90. *Singapore Journal of Tropical Geography, 15*(2), 95–111.

Auty, R. M., & Warhurst, A. (1993). Sustainable development in mineral exporting economies. *Resources Policy, 19*(1), 14–29.

Baena, C., Sevi, B., & Warrack, A. (2012). Funds from non-renewable energy resources: Policy lessons from Alaska and Alberta. *Energy Policy, 51*, 569–577.

Bailey, J. (1996). High seas fishing: Towards a sustainable regime. *Sociologia Ruralis, 36*(2), 189.

Baklanov, P. Y., & Moshkov, A. V. (2016). Structural transformations of the economy in the Pacific region of Russia and efficiency trends. *Ekonomika Regiona-Economy of Region, 12*(1), 46–63.

Bebbington, J., Brown, J., & Frame, B. (2007). Accounting technologies and sustainability assessment models. *Ecological Economics, 61*(2–3), 224–236.

Beckerman, W. (1994). "Sustainable development": Is it a useful concept? *Environmental Values, 3*(3), 191–209.

Belonozhko, M. L., Silin, A. N., & Gyurdzhinyan, A. S. (2018). Scientific accompaniment of forming of sustainable human potential in the oil and gas Arctic region. *International Journal of Ecology & Development, 33*(1), 138–144.

Berardi, G. (1998). Natural resource policy, unforgiving geographies, and persistent poverty in Alaska native villages. *Natural Resources Journal, 38*(1), 85–108.

Bhattacharya, J. (2000). Sustainable development of natural resources: Implications for mining of minerals. *Mineral Resources Engineering, 9*(4), 451–464.

Blair, B., Lovecraft, A. L., & Kofinas, G. P. (2014). Meeting institutional criteria for social resilience: A nested risk system model. *Ecology and Society, 19*(4).

Boiral, O. (2013). Sustainability reports as simulacra? A counter-account of A and A plus GRI reports. *Accounting Auditing & Accountability Journal, 26*(7), 1036–1071.

Bradshaw, M. J. (1995). The Russian North in transition: General introduction. *Post-Soviet Geography, 36*(4), 195–203.

Bradshaw, M. J. (2010). Global energy dilemmas: A geographical perspective. *Geographical Journal, 176*, 275–290.

Caddy, J. F. (1999). Fisheries management in the twenty-first century: Will new paradigms apply? *Reviews in Fish Biology and Fisheries, 9*, 1–43.

Carothers, C. (2008). "Rationalized Out": Discourses and realities of fisheries privatization. In A. I. M. E. L. Kodiak & C. Carothers (Eds.), *Enclosing the fisheries: People, places, and power, 68*, 55–74.

Caulfield, R. A. (1993). Aboriginal subsistence whaling in Greenland: The case of Qeqertarsuaq municipality in West Greenland. *Arctic, 46*(2), 144–155.

Chambers, C., & Kokorsch, M. (2017). Viewpoint: The social dimension in Icelandic fisheries governance. *Coastal Management, 45*(4), 330–337.

Chance, N. A., & Andreeva, E. N. (1995). Sustainability, equity, and natural resource development in Northwest Siberia and Arctic Alaska. *Human Ecology, 23,* 217–240.

Chapin, F. S., Sommerkorn, M., Robards, M. D., & Hillmer-Pegram, K. (2015). Ecosystem stewardship: A resilience framework for arctic conservation. *Global Environmental Change-Human and Policy Dimensions, 34*, 207–217.

Chiou, H. K., Tzeng, G. H., & Cheng, D. C. (2005). Evaluating sustainable fishing development strategies using fuzzy MCDM approach. *Omega-International Journal of Management Science, 33*(3), 223–234.

Clapp, R. A. (1998). The resource cycle in forestry and fishing. *Canadian Geographer-Geographe Canadien, 42*(2), 129–144.

Costa-Pierce, B. A. (2010). Sustainable ecological aquaculture systems: The need for a new social contract for aquaculture development. *Marine Technology Society Journal, 44*(3), 88–112.

Cragg, A. W. (1998). Sustainable development and mining: Opportunity or threat to the industry? *Cim Bulletin, 91*(1023), 45–50.

Dahl, E., Ervik, A., Iversen, S. A., Moksness, E., & Saetre, R. (2008). Reconciling fisheries with conservation in the coastal zones—The Norwegian experience and status. In J. Nielsen, J. J. Dodson, K. Friedland, T. R. Hamon, J. Musick, & E. Verspoor (Eds.), *Reconciling fisheries with conservation*. American Fisheries Society Symposium Series 46, 1519.

Dashwood, H. S. (2014). Sustainable development and industry self-regulation: Developments in the global mining sector. *Business and Society, 53*(4), 551–582.

Davankov, A. Y., Dvinin, D. Y., & Postnikov, Y. A. (2016). Methodological tools for the assessment of ecological and socioeconomic environment in the region within the limits of the sustainability of biosphere. *Ekonomika Regiona-Economy of Region, 12*(4), 1028–1039.

Davis, A., & Jentoft, S. (2001). The challenge and the promise of indigenous peoples' fishing rights—from dependency to agency. *Marine Policy, 25*(3), 223–237.

Davis, B. (1996). Achieving sustainable development: Scientific uncertainty and policy innovation in Tasmanian regional development. *Australian Journal of Public Administration, 55*(4), 100–108.

de Villiers, C., Low, M., & Samkin, G. (2014). The institutionalization of mining company sustainability disclosures. *Journal of Cleaner Production, 84*, 51–58.

Drummond, I., & Symes, D. (1996). Rethinking sustainable fisheries: The realist paradigm. *Sociologia Ruralis, 36*(2), 152.

Dudin, M. N., Lyasnikov, N. V., Protsenko, O. D., & Tsvetkov, V. A. (2017). Quantification and risk assessment of hydrocarbon resources development projects in the Arctic region. *Ekonomicheskaya Politika, 12*(4), 168–195.

Duerden, F. (1992). A critical look at sustainable development in the Canadian North. *Arctic, 45*(3), 219–225.

Ediger, V. S., Hosgor, E., Surmeli, A. N., & Tatlidil, H. (2007). Fossil fuel sustainability index: An application of resource management. *Energy Policy, 35*(5), 2969–2977.

Eglington, A., Israel, R., & Vartanov, R. (1998). Towards sustainable development for the Murmansk region. *Ocean & Coastal Management, 41*(2–3), 257–271.

Emberson-Bain, A. (1994). Mining development in the Pacific: Are we sustaining the unsustainable? In W. Harcourt (Ed.), *Feminist perspectives on sustainable development* (pp. 46–59). London: Zed Books.

Escobar, L. F., & Vredenburg, H. (2011). Multinational oil companies and the adoption of sustainable development: A resource-based and institutional theory interpretation of adoption heterogeneity. *Journal of Business Ethics, 98*(1), 39–65.

Eythorsson, E. (1996). Coastal communities and ITQ management. The case of Icelandic fisheries. *Sociologia Ruralis, 36*(2), 212.

Fauchald, P., Hausner, V. H., Schmidt, J. I., & Clark, D. A. (2017). Transitions of social-ecological subsistence systems in the Arctic. *International Journal of the Commons, 11*(1), 275–329.

Fitzpatrick, P., Fonseca, A., & McAllister, M. L. (2011). From the Whitehorse Mining Initiative towards sustainable mining: Lessons learned. *Journal of Cleaner Production, 19*(4), 376–384.

Fleury, A. M., & Davies, B. (2012). Sustainable supply chains-minerals and sustainable development, going beyond the mine. *Resources Policy, 37*(2), 175–178.

Foley, P., & Havice, E. (2016). The rise of territorial eco-certifications: New politics of transnational sustainability governance in the fishery sector. *Geoforum, 69*, 24–33.

Foley, P., & Hebert, K. (2013). Alternative regimes of transnational environmental certification: Governance, marketization, and place in Alaska's salmon fisheries. *Environment and Planning A-Economy and Space, 45*(11), 2734–2751.

Fonseca, A., McAllister, M. L., & Fitzpatrick, P. (2014). Sustainability reporting among mining corporations: A constructive critique of the GRI approach. *Journal of Cleaner Production, 84*, 70–83.

Forbes, B. C. (2013). Cultural resilience of social-ecological systems in the Nenets and Yamal-Nenets Autonomous Okrugs, Russia: A focus on Reindeer Nomads of the Tundra. *Ecology and Society, 18*(4).

Forbes, B. C., & Stammler, F. (2009). Arctic climate change discourse: The contrasting politics of research agendas in the West and Russia. *Polar Research, 28*(1), 28–42.

Gallardo, G. L., Saunders, F., Sokolova, T., Boreback, K., van Laerhoven, F., Kokko, S., & Tuvenda, M. (2017). We adapt, but is it good or bad? Locating the political ecology and social-ecological systems debate in reindeer herding in the Swedish Sub-Arctic. *Journal of Political Ecology, 24*, 667–691.

Gerlach, S. C., & Loring, P. A. (2013). Rebuilding northern foodsheds, sustainable food systems, community wellbeing, and food security. *International Journal of Circumpolar Health, 72*, 87–90.

Gilman, E., Passfield, K., & Nakamura, K. (2014). Performance of regional fisheries management organizations: Ecosystem-based governance of bycatch and discards. *Fish and Fisheries, 15*(2), 327–351.

Glazyrina, I. P. (1998). Looking for a path to sustainability in Eastern Siberia. *Ecosystem Health, 4*(4), 248–255.

Glen, J. J. (1995). Sustainable yield analysis in a multicohort single-species fishery: A mathematical programming approach. *Journal of the Operational Research Society, 46*(9), 1052–1062.

Gulbrandsen, L. H., & Honneland, G. (2014). Fisheries certification in Russia: The emergence of non-state authority in a post-Communist economy. *Ocean Development and International Law, 45*(4), 341–359.

Gullestad, P., Abotnes, A. M., Bakke, G., Skern-Mauritzen, M., Nedreaas, K., & Sovik, G. (2017). Towards ecosystem-based fisheries management in Norway— Practical tools for keeping track of relevant issues and prioritizing management efforts. *Marine Policy, 77*, 104–110.

Gunn, A., Russell, D., & Greig, L. (2014). Insights into integrating cumulative effects and collaborative co-management for migratory tundra caribou herds in the Northwest Territories, Canada. *Ecology and Society, 19*(4).

Hall, J., Matos, S., & Silvestre, B. (2012). Understanding why firms should invest in sustainable supply chains: A complexity approach. *International Journal of Production Research, 50*(5), 1332–1348.

Hansen, W. D. (2014). Generalizable principles for ecosystem stewardship-based management of social-ecological systems: Lessons learned from Alaska. *Ecology and Society, 19*(4).

Harsem, O., & Hoel, A. H. (2013). Climate change and adaptive capacity in fisheries management: The case of Norway. *International Environmental Agreements-Politics Law and Economics, 13*(1), 49–63.

Hebert, K. (2015). Enduring capitalism: Instability, precariousness, and cycles of change in an Alaskan Salmon fishery. *American Anthropologist, 117*(1), 32–46. https://doi.org/doi:10.1111/aman.12172

Hilborn, R. (2010). Pretty good yield and exploited fishes. *Marine Policy, 34*(1), 193–196.

Hilborn, R., Parrish, J. K., & Litle, K. (2005). Fishing rights or fishing wrongs? *Reviews in Fish Biology and Fisheries, 15*, 191–199.

Hilson, G., & Murck, B. (2000). Sustainable development in the mining industry: Clarifying the corporate perspective. *Resources Policy, 26*(4), 227–238.

Himes-Cornell, A., & Hoelting, K. (2015). Resilience strategies in the face of short- and long-term change: Out-migration and fisheries regulation in Alaskan fishing communities. *Ecology and Society, 20*(2).

Horsley, J., Prout, S., Tonts, M., & Ali, S. H. (2015). Sustainable livelihoods and indicators for regional development in mining economies. *Extractive Industries and Society-an International Journal, 2*(2), 368–380.

Humphreys, D. (2001). Sustainable development: Can the mining industry afford it? *Resources Policy, 27*(1), 1–7.

Huntington, H. (2014). *Traditional knowledge and resource development* [Gap Analysis #11]. Retrieved from Yukon College website: http://yukonresearch. yukoncollege.yk.ca/wpmu/wp-content/uploads/sites/2/2013/09/11-ReSDA-Huntington-TK-Final.pdf

Huskey, L., & Southcott, C. (2016). "That's where my money goes": Resource production and financial flows in the Yukon economy. *The Polar Journal, 6*(1), 11–29.

Johannesson, S. E., Daviosdottir, B., & Heinonen, J. T. (2018). Standard ecological footprint method for small, highly specialized economies. *Ecological Economics, 146*, 370–380.

Kadenic, M. D. (2015). Socioeconomic value creation and the role of local participation in large-scale mining projects in the Arctic. *Extractive Industries and Society-an International Journal, 2*(3), 562–571.

Klokov, K., & Khrushchev, S. (2016). Comparative ethno-ecological evaluation of the sustainability of traditional reindeer herding on Kola and Yamal Peninsulas. *SGEM, 3*, 31–38.

Korobitsyn, B. A. (2015). Methodological approaches for estimating gross regional product after taking into account depletion of natural resources, environmental pollution and human capital aspects. *Ekonomika Regiona-Economy of Region, 3*, 77–88.

Kryukov, V. A., Sevastyanova, A. Y., Tokarev, A. N., & Shmat, V. V. (2017). A modern approach to the elaboration and selection of strategic alternatives for resource regions. *Ekonomika Regiona-Economy of Region, 13*(1), 93–105.

Ksenofontov, S., Backhaus, N., & Schaepman-Strub, G. (2017). "To fish or not to fish?": Fishing communities of Arctic Yakutia in the face of environmental change and political transformations. *Polar Record, 53*(3), 289–303.

Kyllonen, S., Colpaert, A., Heikkinen, H., Jokinen, M., Kumpula, J., Marttunen, M., & Raitio, K. (2006). Conflict management as a means to the sustainable use of natural resources. *Silva Fennica, 40*(4), 687–728.

Lahiri-Dutt, K. (Ed.). (2011). *Gendering the field: Towards sustainable livelihoods for mining communities*. Canberra: ANU R Press.

Langdon, S. J. (2008). The community quota program in the Gulf of Alaska: A vehicle for Alaska native village sustainability? In M. E. Lowe & C. Carothers (Eds.), *Enclosing the fisheries: People, places, and power* (pp. 155–194).

Lertzman, D., & Vredenburg, H. (2005). Indigenous peoples, resource extraction and sustainable development: An ethical approach. *Journal of Business Ethics, 56*(3), 239–254.

Lindahl, K. B., Johansson, A., Zachrisson, A., & Viklund, R. (2018). Competing pathways to sustainability? Exploring conflicts over mine establishments in the Swedish mountain region. *Journal of Environmental Management, 218*, 402–415.

Loring, P. A. (2017). The political ecology of gear bans in two fisheries: Florida's net ban and Alaska's Salmon wars. *Fish and Fisheries, 18*(1), 94–104.

Low, N., & Gleeson, B. (1998). Situating justice in the environment: The case of BHP at the Ok Tedi Copper Mine. *Antipode*. https://doi.org/doi:10.1111/1467-8330.00075

Lydall, M. (2009). Backward linkage development in the South African PGM industry: A case study. *Resources Policy, 34*(3), 112–120.

Lyons, C., Carothers, C., & Reedy, K. (2016). A tale of two communities: Using relational place-making to examine fisheries policy in the Pribilof Island communities of St. George and St. Paul, Alaska. *Maritime Studies, 15*, 7. doi:10.1186/s40152-016-0045-1.

MacKenzie, J. J. (1997). Two transport visions. *Annals of the American Academy of Political and Social Science, 553*, 192–198.

Mangel, M. (2010). Scientific inference and experiment in Ecosystem Based Fishery Management, with application to Steller sea lions in the Bering Sea and Western Gulf of Alaska. *Marine Policy, 34*(5), 836–843.

McGoodwin, J. R. (2007). Effects of climatic variability on three fishing economies in high-latitude regions: Implications for fisheries policies. *Marine Policy, 31*(1), 40–55.

Moffat, K., Lacey, J., Zhang, A., & Leipold, S. (2016). The social license to operate: A critical review. *Forestry, 89*(5), 477–488.

Munk-Madsen, E. (1998). The Norwegian fishing quota system: Another patriarchal construction? *Society & Natural Resources, 11*(3), 229–240.

Natcher, D., Shirley, S., Rodon, T., & Southcott, C. (2016). Constraints to wildlife harvesting among aboriginal communities in Alaska and Canada. *Food Security, 8*(6), 1153–1167.

Nelsen, J. L., Scoble, M., & Ostry, A. (2010). Sustainable socioeconomic development in mining communities: North-central British Columbia perspectives. *International Journal of Mining Reclamation and Environment, 24*(2), 163–179.

Nikolakis, W., Nelson, H. W., & Cohen, D. H. (2014). Who pays attention to indigenous peoples in sustainable development and why? Evidence from socially responsible investment mutual funds in North America. *Organization & Environment, 27*(4), 368–382.

Novoselov, A., Potravny, I., Novoselova, I., & Gassiy, V. (2017). Selection of priority investment projects for the development of the Russian Arctic. *Polar Science, 14*, 68–77.

Nygaard, V. (2016). Do indigenous interests have a say in planning of new mining projects? Experiences from Finnmark, Norway. *Extractive Industries and Society*, *3*(1), 17–24.

Otterstad, O. (1996). Sustainable development in fisheries: Illusions or emerging reality? *Sociologia Ruralis*, *36*(2), 163. https://doi.org/doi:10.1111/j.1467-9523.1996.tb00013.x

Parlee, B., Sandlos, J., & Natcher, D. (2018). Undermining subsistence: Barren-ground caribou in a "tragedy of open access." *Science Advances*, *4*(2).

Pelyasov, A. N., Galtseva, N. V., & Atamanova, E. A. (2017). Economy of the Arctic "Islands": The case of Nenets and Chukotka Autonomous Okrugs. *Ekonomika Regiona-Economy of Region*, *13*(1), 114–125.

Pettersen, J. B., & Song, X. Q. (2017). Life cycle impact assessment in the Arctic: Challenges and research needs. *Sustainability*, *9*(9).

Prior, T., Giurco, D., Mudd, G., Mason, L., & Behrisch, J. (2012). Resource depletion, peak minerals and the implications for sustainable resource management. *Global Environmental Change-Human and Policy Dimensions*, *22*(3), 577–587.

Prno, J. (2013). An analysis of factors leading to the establishment of a social license to operate in the mining industry. *Resources Policy*, *38*(4), 577–590.

Prno, J., & Slocombe, D. S. (2012). Exploring the origins of "social license to operate" in the mining sector: Perspectives from governance and sustainability theories. *Resources Policy*, *37*(3), 346–357.

Ramirez-Cendrero, J. M., & Wirth, E. (2016). Is the Norwegian model exportable to combat Dutch disease? *Resources Policy*, *48*, 85–96.

Ray, L. A., Kolden, C. A., & Chapin, F. S. (2012). A case for developing place-based fire management strategies from traditional ecological knowledge. *Ecology and Society*, *17*(3).

Ray, S. (2008). A case study of Shell at Sakhalin: Having a whale of a time? *Corporate Social Responsibility and Environmental Management*, *15*(3), 173–185.

Reedy-Maschner, K. L., & Maschner, H. D. G. (2013). Sustaining Sanak Island, Alaska: A cultural land trust. *Sustainability*, *5*(10), 4406–4427.

Riabova, L., & Didyk, V. (2015). Social license to operate for the resource extraction companies as a new instrument of municipal development. *Voprosy Gosudarstvennogo I Munitsipalnogo Upravleniya–Public Administration Issues, 3*, 61–82.

Richards, J. P. (Ed.). (2009). *Mining, society and a sustainable world.* Heidelberg: Springer.

Robards, M. D., & Greenberg, J. A. (2007). Global constraints on rural fishing communities: Whose resilience is it anyway? *Fish and Fisheries*, *8*(1), 14–30.

Rodon, T., Levesque, F., & Blais, J. (2013). De Rankin Inlet à Raglan, le développement minier et les communautés inuit Etudes inuit. *Inuit studies*, *37*(2), 103–122.

Rodon, T., & Schott, S. (2014). Towards a sustainable future for Nunavik. *Polar Record*, *50*(3), 260–276.

Sachs, J. D., & Warner, A. M. (2001). The curse of natural resources. *European Economic Review*, *45*(4–6), 827–838.

Sairinen, R., Tiainen, H., & Mononen, T. (2017). Talvivaara mine and water pollution: An analysis of mining conflict in Finland. *Extractive Industries and Society*, *4*(3), 640–651.

Salmi, O. (2008). Drivers for adopting environmental management systems in the post-Soviet mining industry. *International Environmental Agreements-Politics Law and Economics, 8*(1), 51–77.

Shvarts, E. A., Pakhalov, A. M., & Knizhnikov, A. Y. (2016). Assessment of environmental responsibility of oil and gas companies in Russia: The rating method. *Journal of Cleaner Production, 127,* 143–151.

Southcott, C., Abele, F., Natcher, D., & Parlee, B. (Eds.). (2018). *Resources and sustainable development in the Arctic.* London: Routledge.

Southcott, C., Walker, V., Wilman, J., Spavor, C., & MacKenzie, K. (n.d.). *The Social Economy and Nunavut: Barriers and Opportunities* (No. 1; SERNNoCa Research Report Series). Northern Research Institute. http://yukonresearch.yukoncollege .yk.ca/frontier/files/sernnoca/PortraitureReportforNunavutv.pdf

Southcott, C., & Natcher, D. (2018). Extractive industries and Indigenous subsistence economies: A complex and unresolved relationship. *Canadian Journal of Development Studies,* 1–18.

Stetson, G., & Mumme, S. (2016). Sustainable development in the Bering Strait: Indigenous values and the challenge of collaborative governance. *Society & Natural Resources, 29*(7), 791–806.

Stiglitz, J. E. (2007). What is the role of the State? In M. Humphreys, J. Sachs, & J. E. Stiglitz (Eds.), *Escaping the resource curse* (pp. 23–52). New York, NY: Columbia University Press.

Suutarinen, T. (2015). Local natural resource curse and sustainable socioeconomic development in a Russian mining community of Kovdor. *Fennia-International Journal of Geography, 193*(1), 99–116.

Symes, D., & Crean, K. (1995). Privatization of the commons: The introduction of individual transferable quotas in developed fisheries. *Geoforum, 26*(2), 175–185.

Thomson, I., & Joyce, S. (2006). Changing mineral exploration industry approaches to sustainability. In M. D. Doggett & J. R. Parry (Eds.), *Wealth creation in the minerals industry: Integrating science, business, and education* (pp. 149–169).

Thornton, T. F. (2015). The ideology and practice of Pacific Herring cultivation among the Tlingit and Haida. *Human Ecology, 43*(2), 213–223.

Thornton, T. F., & Hebert, J. (2015). Neoliberal and neo-communal herring fisheries in Southeast Alaska: Reframing sustainability in marine ecosystems. *Marine Policy, 61,* 366–375.

Tiainen, H. (2016). Contemplating governance for social sustainability in mining in Greenland. *Resources Policy, 49,* 282–289.

Tulaeva, S., & Tysiachniouk, M. (2017). Benefit-sharing arrangements between oil companies and Indigenous people in Russian Northern regions. *Sustainability, 9*(8).

Tynkkynen, V. P. (2007). Resource curse contested: Environmental constructions in the Russian periphery and sustainable development. *European Planning Studies, 15*(6), 853–870.

Tysiachniouk, M. S., & Petrov, A. N. (2018). Benefit sharing in the Arctic energy sector: Perspectives on corporate policies and practices in Northern Russia and Alaska. *Energy Research & Social Science, 39,* 29–34.

Uwasu, M., & Yabar, H. (2011). Assessment of sustainable development based on the capital approach. *Ecological Indicators, 11*(2), 348–352.

van Bets, L. K. J., van Tatenhove, J. P. M., & Mol, A. P. J. (2016). Liquefied natural gas production at Hammerfest: A transforming marine community. *Marine Policy, 69*, 52–61.

Ween, G. B., & Colombi, B. J. (2013). Two rivers: The politics of wild Salmon, Indigenous rights and natural resource management. *Sustainability, 5*(2), 478–495.

Wennecke, C. (2017). Political-economic indicators for self-sustainability in Greenland. *Northern, 45*, 93–111.

Wilson, E., & Stammler, F. (2016). Beyond extractivism and alternative cosmologies: Arctic communities and extractive industries in uncertain times. *Extractive Industries and Society-an International Journal, 3*(1), 1–8.

Wingard, J. D. (2000). Community transferable quotas: Internalizing externalities and minimizing social impacts of fisheries management. *Human Organization, 59*(1), 48–57.

5 Governance for Arctic sustainability

Gary N. Wilson, Gail Fondahl, and Klaus Georg Hansen

Introduction

If sustainable development is a process, it is also a governance challenge. Good governance is required to guide the process and steer institutions and practices. The Arctic is home to a complex, interconnected web of local, regional, national, and global governmental and non-governmental institutions, including Indigenous governance initiatives, many of which aim to facilitate sustainability. The multiplicity of governance complicates efforts to achieve sustainable development. We first define governance and then review briefly conceptual perspectives about sustainability governance. We examine research trends for the Arctic and consider key focus areas, concluding with identification of key research areas.

Definitions and meanings

Although sustainable development, sustainability, and governance are ubiquitous in climate change and environmental discussion, critics charge that they are conceptually ambiguous, without clear definitions (Williams & Millington, 2004; Lange, Driessen, Sauer, Borenmann, & Burger, 2013). Indeed, Jordan (2008) writes "in 'sustainable development' and 'governance' we possibly have two of the most essentially contested terms in the entire social sciences" (p. 18).

However, imprecision is increasingly considered an asset (Jordan, 2008; Veld, 2013; Baker, 2016), and understandings of all three concepts are context dependent and dynamic. This chapter adopts the definitions of sustainability and sustainable development in the Introduction and focuses on the concept of governance and its role in promoting sustainable development.

According to Meadowcroft and Bregha (2009, p. 1): "Governance is "the set of processes through which societies are governed: it includes the actions of government but also those of other societal actors in so far as they contribute to ordering social interactions." As van Zeijl-Rozema, Cövers, Kemp, and Martens (2008, p. 411) note: "It is a collection of rules, stakeholder involvement and processes to realize a common goal." Young (2013, p. 269) defines governance as "a social function centered on steering societies toward socially desirable outcomes and away from socially undesirable outcomes."

Therefore, governance includes the purposeful actions of individuals and non-state groups and traditional, top-down intervention and administration by governments to address societal issues. Shift from a focus on governments to governance acknowledges the inclusion in policy development and decision making of non-governmental actors for multiple reasons. First, there is growing acknowledgement that sustainability issues are multi-faceted, multi-scalar, cross-sectoral, and complex (Shiroyama et al., 2012). Solutions require involving multiple actors and representing different stakeholders. Many crucial issues are transboundary issues that reach beyond governmental jurisdictions in mandates and geographical scale and demand participatory governance designs that include actors inside and outside government. Diverse backgrounds, knowledge, values, and viewpoints can inform solutions and contribute flexibility and resilience for sustainable development (Shiroyama et al., 2012). Recognizing the benefits of diversity includes government and non-governmental actors, and while such diversity and representation are welcome, they burden small, remote, and capacity-challenged governments and non-governmental organizations (NGOs).

Second, expanding governance beyond governments raises expectations regarding participation by non-traditional actors in addressing societal challenges. The public legitimacy of decision making rests on governance processes that involve multiple stakeholders (Griffin, 2010). Sharing responsibility for policymaking can improve its effectiveness and public legitimacy (Lawhon & Patel, 2013). The role of governments at the national and sub-national levels may still be critical, especially for providing structures (political and legal) through which collective goals may be identified and enabled and in making resources available to support collective actions to achieve such goals (Meadowcroft & Bregha, 2009). But non-state actors frequently claim entitlement to participate alongside governments in policy development and decision making.

Evolving conceptual perspectives on governance for Arctic sustainability

Trends in conceptualizing governance for sustainability

Recent focus has refined the concept, including multilevel governance (Newig & Fritsch, 2009; Koivurova, 2013; Hooghe & Marks, 2003),

reflexive governance (Voss & Borenmann, 2011), interactive governance (Kooiman, Bavinck, Chuenpadgee, Mahon, & Pullin, 2008), adaptive governance (Karpouzoglou, Dewulf, & Clark, 2016; Folke, Hahn,
Olsson, & Norberg, 2005; Kofinas, 2009), and trans-governance (Veld,
2013). More specifically, governance for sustainability is captured under
an array of terms: environmental (Glasbergen, 1998; Newig & Fritsch,
2009; Paterson, Humphreys, & Pettiford, 2003; Duyck, 2012, 2015; Stossel, Tedsen, Cavalieri, & Riedel, 2014), earth systems (Biermann, 2007),
socio-ecological (Kofinas, 2009), and governance for resilience (Huitric
et al., 2009). Employing searches of these terms on GoogleScholar[1] as
a proxy to examine trends perspectives on governance for sustainability
over the past 25 years reveals increased attention to the topic generally and
in the Arctic (Figure 5.1). For the general literature, a jump takes place in
1997 and for both the rate of growth increases after 2007. The latter may
indicate connection to the International Polar Year (2007–2008).

There is increasing recent tendency to view governance through a socio-
ecological lens generally and in the Arctic literature (Figure 5.2). Conceptualizing the Arctic as a dynamic social-ecological system (SES) has gained
significant traction over the past decade (ARR, 2016) and includes pay-
ing attention to governance issues integral to this conception of the Arctic.

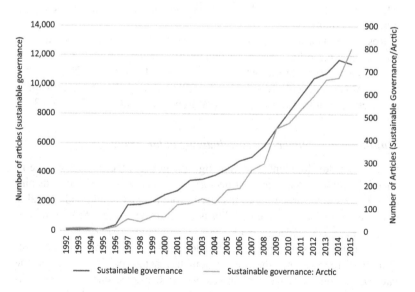

Figure 5.1 Articles on governance for sustainability—general and Arctic. This figure shows results for all terms as noted in note 1.

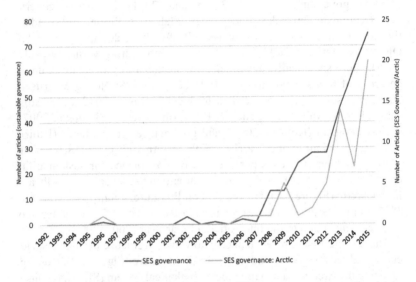

Figure 5.2 Articles on social-ecological system (SES) governance—general and Arctic. This figure shows results for the searched terms 'social-ecological,' 'socio-ecological governance,' and 'SES' with governance.

Although the number of SES articles is relatively small, they have grown quickly in number.

Resilience as a framing concept has also penetrated the governance literature, as it has other areas of sustainability science (Beunen, Patterson, & Van Assche, 2017; Garmestani & Benson, 2013; Petrov et al., 2016; Smith, 2014). This is particularly notable after 2009, and uptake of this concept in the Arctic literature follows the general trends for resilience (Figure 5.3).

Multi-level governance and sustainability in the Arctic

Sustainable development is a challenge for governance, meaning governance is necessary to direct the trajectory of collective actions—the sustainable development process—to foster environmentally sound and equitable outcomes. The lack of sustainability is often attributed to a failure of governance (van Zeijl-Rozema et al., 2008; Lange et al., 2013).

Sustainability issues were mostly addressed by governmental organizations until recently, but the role of non-state actors began to grow in the late 20th century. International processes and decisions, such as the Brundtland report (1987), drive this shift. Similarly, the 1992 Rio Declaration on Environment and Development noted that "[e]nvironmental issues are best

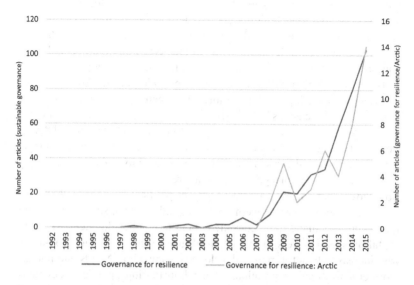

Figure 5.3 Articles on resilience governance—general and Arctic. This figure
shows results for the searched terms 'resilience' and 'resilience-based'
with governance.

handled with participation of all concerned citizens, at the relevant level"
(UNEP, 1992, Principle 10). Expansion of the role of non-state actors in governance, for the purpose of sustainable development, was caused by an ideo-
logical shift in Western democracies aimed at reducing the role of the state
and empowering non-state actors. This neo-liberal shift "rolled back" the
power of governments and the state and reinforced this process by introducing new processes and institutions, such as free trade agreements and public–
private partnerships (P3s) (Young & Matthews, 2007; Peck & Tickell, 2002).

Lange et al. (2013, p. 406) call sustainable development a "paradigm
case of what motivated the governance approach in general: the increasingly complex, dynamic and interdependent nature of contemporary policy
making." Hierarchical governing by formal governments is inadequate to
address the "wicked" problems of significant social, environmental, and
political complexity that sustainable development poses. Managing diverse
values, understandings, and perspectives regarding sustainability to achieve
consensus demands participatory governance with multiple actors (van
Zeijl-Rozema et al., 2008). Requisites for pursuing sustainable development
include "a climate of 'dialogue', 'partnership' and 'shared responsibility',
rather than centralized 'command and control' policies'" with "stakeholder
participation and partnership between policy levels, institutions and actors"

(Griffin, 2010, p. 366). The literature on governance for sustainability underscores involving multiple civil society actors so that governance is collaborative, polycentric, reflexive, and adaptive (Lange et al., 2013). It recognizes that multiple actors bring various types of expertise and information to the table, positively contributing to policymaking processes.

The literature on multilevel governance provides a useful framework for analyzing relationships within complex governance regimes, including the Arctic. Two "dimensions" of governance are highlighted, vertical and horizontal (Poelzer & Wilson, 2014; Wilson, 2017) in which the vertical dimension refers to governments and government-based institutions arranged along a local–global vertical continuum from local to global. Traditionally, national and sub-national governments are responsible for policies and decisions governing society and are critical in political-economic administration. More recently, new governments—including local, supranational, and Indigenous—demand to be consulted about policy development. The horizontal dimension refers to expansion of governance to include non-actors who increasingly act in policy development, including NGOs and not-for-profit organizations, corporations, research institutions, public associations, and other civil society actors. Horizontal expansion of governance is encouraged by neo-liberal policies and by societal shifts that empower citizens and raise their expectations about consultation and political involvement. While this represents a positive political development for democracy and citizen engagement, it also complicates and lengthens decision-making processes by requiring enhanced collaboration. Such collaboration can lead to "lowest common denominator" solutions that appeal to the widest variety of stakeholders rather than targeted, focused policies that address real issues.

The Arctic is characterized by complex, dynamic systems of governmental and non-governmental institutions spanning the political spectrum from the local to circumpolar scale. It includes long-established settler governments, more recently established Indigenous governments, and supranational and international governance bodies that function like governments. Each governance level, alongside institutions that operate outside government, is involved in sustainable development, and its vertical and horizontal dimensions influence it.

Governance for Arctic sustainability: current critical themes and issues

The evolution of innovative governance structures and their contributions to sustainability in the Arctic

Geopolitically, Young (2010) has argued that a 'socio-ecological state change' in the Arctic in the latter 20th century has been followed by a more recent state change, both with implications for governance for sustainability. The first state change, a product of the end of the Cold War and framing of the Arctic as a region of 'peace' and 'cooperation,' enabled innovative regional and circumpolar governance initiatives. As a result, the Arctic became recognized as a region with distinct governance challenges and opportunities. Recognizing the transboundary nature of environmental challenges further encouraged international cooperation, instigating innovative inter-governmental cooperation on sustainability agendas. Lower levels of government have also experienced governance innovations, especially those that have empowered Indigenous peoples.

More recent changes in biophysical processes (climatic warming) and globalization (resource development, circumpolar transportation routes) suggest the need for governance across and beyond the Arctic (Young, 2010; Keil & Knecht, 2017). Inuit leader and activist Sheila Watt-Cloutier observes that while the Arctic is disproportionately affected by climate change, it serves as a "canary in the coalmine," warning about the future implications for other regions (2015). Next we discuss the Arctic Council and how it encourages interaction between Arctic states and Indigenous peoples.

Inter-governmental forum: the Arctic Council

The Arctic Environmental Protection Strategy (AEPS), founded in 1991, and the Arctic Coucil, founded in 1996, are pivotal developments in Arctic governance for sustainability at the circumpolar scale. The Arctic Council and its working groups and subsidiary bodies connect the eight Arctic states, six Arctic Indigenous organizations, and non-Arctic observer states and organizations in discussions about regional environment and human development.

For over two decades, the Council has been "the leading intergovernmental forum promoting cooperation, coordination and interaction among the Arctic States, Arctic Indigenous communities and other Arctic inhabitants on issues [such as] sustainable development and environmental protection" (Arctic Council, 2017a). Discussions of the Arctic Council as a governance institution frequently focus on key governance issues, including Indigenous actors in Council deliberations and increased use of Indigenous knowledge and Indigenous governance protocols (Koivurova, 2011); the limited but growing scope of its governance activities (Kankaanpää & Young, 2012); the legitimacy of state actors and organizations outside the Arctic (Graczyk & Koivurova, 2014); and competing governance initiatives that may weaken the Council's ability to promote sustainability (Exner-Pirot & Plouffe, 2013).

The Arctic Council's mandate is limited. First, like other international bodies such as the United Nations, its ability to enforce decisions is limited by decision-making capacities of state governments. The scope of the Council's policy purview is restricted to "soft" security matters such as environmental and social security; the Council's Charter prevents it from discussing "hard" security or military matters. Discussions about soft security can be controversial because there has been a perceived shift from environmental security to human development social and economic aspects of sustainability and resource development (Koivurova, 2010). This shift triggers concern about the Council's focus on environmental sustainability (Tesar, Dubois, Sommerkorn, & Shetakov, 2016; Conley & Melino, 2016). Moreover, concerns exist about the Council's ability to pursue a sustainability agenda because it cannot compel member states to act (Graczyk & Koivurova, 2014; Young, 2010; Wehrmann, 2016). The Council's unity was challenged by the Ilulissat Declaration of May 2008 by the Arctic Five (Koivurova, 2011; Young, 2010),[2] the subset of Arctic states bordering the Arctic Ocean and that have convened separately to discuss Arctic Ocean governance issues (Kuersten, 2016).

Rights holders: the expansion of indigenous roles in Arctic governance for sustainability

Increasing Indigenous participation in governance, at all levels, has occurred over the last few decades. The Arctic Council includes Arctic Indigenous organizations in Council deliberations, and as "Permanent Participants," six Indigenous organizations[3] have full consultation rights in deliberations and meetings (Wilson, 2017)—albeit more limited than those of the member states and not always satisfying some Indigenous organizations (Koivurova, 2010, 2011).

International Arctic Indigenous organizations are also important in promoting regional cooperation on sustainability governance issues. The Inuit Circumpolar Council, representing Inuit across four countries, works with several United Nations organizations on issues of climate change, cultural sustainability, and economic development (ICC Canada, 2016; Shadian, 2014). The Saami Council, representing Saami in four countries in Fennoscandia, works with United Nations organizations and regional organizations, such as the Barents Euro-Arctic Council (Saami Council, 2017).

The emergence of Indigenous self-governing regions over the past 40 years is an important governance development at the national and sub-national levels. Processes such as devolution in Canada (and in 1990s Russia), home rule and self-rule in Greenland, and decentralization in the Nordic countries have empowered regional and local governments with influence over sustainability issues (Poelzer & Wilson, 2014; Fondahl & Wilson, 2017). The emergence of Indigenous regional governments and

Indigenous legislatures, such as the Sami Parliaments in the Nordic countries, introduce new environmental stewardship and preserve Indigenous cultures and languages based on Indigenous philosophies. Addressing conflicting priorities among environmental, economic, and social sustainability will remain important (Bjørst, 2017; Gad, Jakobsen, & Strandsberg, 2017).

Changes brought by decentralization, devolution, and self-government add additional layers to the political system, thereby expanding the vertical dimension of multilevel governance. New opportunities for sustainability have arisen on the horizontal dimension of multilevel governance. For example, the recently established Inuit-Crown Partnership Committee (ICPC) in Canada gathers senior ministers from the federal government and Inuit Tapiriit Kanatami (ITK), an NGO representing Inuit to discuss "shared priority areas" such as housing, education, health, and the environment (Indigenous and Northern Affairs Canada, 2017; Wilson & Selle, 2019). ITK provides Inuit with direct, regular communication with government to influence policy development.

Another example of Arctic governance innovation is co-management, instituted to aid management of renewable and non-renewable resources such as fisheries, parks, wildlife, and plants (White, 2006, Kocho-Schellenberg & Berkes, 2015; Hermann & Martin, 2016). Executed by boards, co-management includes representatives from various levels of government and non-governmental stakeholders and thus combines Indigenous and Western governance processes. Developing governance models that equitably incorporate Indigenous knowledge, processes, and protocols with Western scientific knowledge and protocols presents challenges, and co-management is not always harmonious (Smith, 2013). Yet it remains one of the best ways to bridge Indigenous and Western traditions in a single governance process, and recent research stresses the need for *adaptive* co-management given the rapidly changing biophysical environment (Plummer, Armitage, & de Loë, 2013, Plummer & Baird, 2013).

Governance innovation endogenous to the Arctic includes developing structures to encourage Indigenous economic sustainability. One example, which expands the horizontal dimensions of multilevel governance, are Inuit Economic Development Corporations (IEDC) (Wilson & Alcantara, 2012), which manage and invest funds obtained from land claims agreements on behalf of Inuit beneficiaries. Like regular corporations, IECDs invest in businesses and other ventures to grow the funds under their control. In addition to seeking a monetary return on investments, IEDCs often invest in regional businesses. For example, the Makivik Corporation in Nunavik (northern Québec) owns and operates regional airlines and other transportation enterprises linking communities in this region with the rest of Canada. It contributes to the social and cultural sustainability by investing in social and community programming and housing development (Makivik Corporation, 2017).

In the Sakha Republic (Yakutia) of Russia, municipal administrations with predominantly Indigenous populations have declared themselves as 'national' (i.e., Indigenous) administrations. This status achieves various sustainability goals, such as increased protection against or opportunities for compensation from damage from industrial development. Laws passed at the republican (regional) level make this possible, but they may be compromised by contradictory federal laws, which take precedent over regional ones (Fondahl et al., 2019; Ivanova & Stammler, 2017).

Stakeholders: the question of legitimate participation in Arctic governance

Governance for sustainability is confronted by the rapidly changing environmental situation in the Arctic but also by increased interest from outside the Arctic. Questions such as who should participate in governance decisions, how should that participation be weighted, and whose knowledge is relevant and necessary are asked with increasing urgency. The number of actors who consider themselves to have legitimate interests in the Arctic and Arctic policymaking has risen notably in the past decade at various scales of governance.

A recent marked increase in applications by states and organizations, non-governmental and inter-governmental, for 'observer status' on the Arctic Council is striking (Arctic Council, 2017). Observers' opportunities for input are circumscribed, compared with the Arctic member states and Permanent Participants, but there is concern about motivations (Humbrich, 2013), especially for sustainability agendas. Many recently appointed observer countries, such as China and South Korea, are interested in resource development and transportation opportunities that could jeopardize sustainable use and development of the Arctic (Bennett, 2015; Chaturvedi, 2013; Hong, 2014; Hsiung, 2016; Solli, Wilson Rowe, & Lindgren, 2013). In northern Alaska, communities view incursions by non-Arctic interests as threats to local ecosystems and traditional economies. However, in Greenland, such development is welcomed by national and local governments, which remain open to non-Arctic development and utilization of non-renewable resources.

Specific Arctic governance regimes on scientific cooperation have also been faulted for being too limited in their breadth. The recent Task Force on Enhancing Scientific Cooperation in the Arctic (SCTF, under the aegis of Arctic Council) has been criticized for only benefiting the member states of the Arctic Council (Shibata & Raita, 2016), and the Agreement on Enhancing International Arctic Scientific Cooperation accused of insufficiently dealing with the interests of non-Arctic researchers and of Permanent Participants (e.g., Liu, 2018). Non-Arctic states have launched significant research initiatives Arctic sustainability (Japan's Arctic Challenge for Sustainability project; www.arcs-pro.jp) and have pursued active involvement in international Arctic science organizations, focused on improving sustainability, research, and education in the Arctic, such as the International Arctic Science Committee (IASC), the International Arctic Social Sciences Association (IASSA), and the University of the Arctic (UArctic).

Groups considering themselves as having legitimate but restricted interests in the Arctic may choose alternative paths to influence policy decisions. For example, establishment of the Arctic Circle, a "network of international dialogue and cooperation on the future of the Arctic," was initially seen by some as a direct challenge to the exclusivity of the Arctic Council. One of its recent forums (2016) focused specifically on Arctic sustainable development, including governance issues. Creating more opportunities for stakeholders to link to, but not directly participate in, key Arctic governance bodies such as the Arctic Council have been suggested to ensure wider inclusion in Arctic sustainability discussions (Humbrich, 2013; Young & Kim, 2012). After all, as Watt-Cloutier (2015) noted, decisions made outside the Arctic on issues including industrialization, resource development, and the use of chemicals and pesticides negatively impact the Arctic environment. Greater understanding on the part of non-Arctic states about the consequences of their actions could lead to greater sustainability in the region.

Arctic (terrestrial) natural resource management

Resource development projects represent an important source of income and financial autonomy from higher levels of government and a way to improve wellbeing and economic sustainability. As existing resources in accessible locations are depleted and technological advancement continues, extractive industries become increasingly important, and external demand for resources will influence Arctic sustainability. Many Arctic residents, including Indigenous peoples, support natural resource development, usually because they influence the scale and pace of development and people receive economic benefits. For instance, ICC Canada stated

> Managed under Inuit Nunaat governance structures, non-renewable resource development can contribute to Inuit economic and social development through both private sector channels (employment, income, businesses) and public sector channels (revenues from publicly owned land, tax revenues, infrastructure.
>
> (ICC Canada, n.d.)

Resource development projects, however, can be controversial when they lead to local environmental degradation and political divisiveness. There is no consensus on the best way to balance resource development and environmental sustainability (Forbes & Kofinas, 2014). However, meaningful engagement with local communities from project inception is key.

Governing the Arctic Ocean for sustainability in the face of climate change

The retreat of Arctic sea ice stimulates interest in resource extraction and shipping in the Arctic Ocean. Next we consider the evolving issues of Arctic Ocean governance regarding sustainability, noting recent developments.

Arctic Ocean resource governance

Arctic Ocean resources include oil, gas, fish, and tourism. Under the United Nations Convention on the Law of the Sea (UNCLOS), four Arctic countries (Russia, Canada, Denmark, and Norway) have established territorial claims beyond their 200-nautical-mile exclusive economic zones, largely because of the potential for hydrocarbon resource development. While the United States has yet to ratify UNCLOS, it confirmed its commitment to adhere to the law of the sea through the 2008 Illulissat Declaration (Brosnan, Leschine, & Miles, 2011). Once ratified, such claims could facilitate resource development, thereby jeopardizing sustainable development. The UNCLOS process is an indication of the power that states still wield in Arctic governance despite the emergence of bodies such as the Arctic Council. Developing resources requires enormous investment, which in turn demands geopolitical stability (Brosnan et al., 2011). Environmental concerns about oil spills on the fragile Arctic ecology and the inability to manage these spills, given the remoteness of this region, deters development in parts of the Arctic Ocean.

Concern over the sustainability of fishing stocks as the Arctic Ocean warms has resulted in remarkable governance innovation. In 2018, nine states and the European Union suspended fishing in the High Arctic, committing to further research on fishing stocks (Schatz, Proelss, & Liu, 2019). This signals an interest in a 'knowledge before action' approach, or dealing with sustainable resource use prior to initiating resource exploitation.

In Canada, a recent Supreme Court ruling about acoustic pollution from seismic testing during hydrocarbon exploration found Canada's National Energy Board guilty of a flawed consultation process that failed to adequately consider Inuit rights and subsistence needs (Tasker, 2017).

Requiring greater local stakeholder participation by in governance processes that affect communities is a clear lesson of this ruling.

Managing fishing stocks provides another focus of Arctic governance (Fluharty, 2012; Molenaar, 2014). Of key importance to the national economies of Greenland and Iceland; fisheries are affected by transboundary pollution and overfishing yet critical to local subsistence economies and the socio-cultural health of many Arctic communities. The impact of climate change on fish stocks is uncertain but may involve the migration of marine populations both within and from beyond the Arctic. Governance mechanisms to address Arctic fishing include bilateral and multi-lateral conventions and treaties.

Arctic cruise tourism is a burgeoning activity with significant environmental, economic, and social sustainability dimensions (Dawson, Johnston, & Stewart, 2014; Hall, James, & Wilson, 2010; Stonehouse & Snyder, 2010). Governance occurs partially by the Association of Arctic Expedition Tour Operators (AECO), which voluntary develops Arctic tourism guidelines (Paskevich, Dawson, & Stewart, 2015). Like resource development, however, cruise tourism promises much-needed economic opportunities and revenues, but tourists can overwhelm small communities and the environment. Regional and national governments must provide communities with the resources and training to respond. Likewise, private tour operators must liaise closely with local stakeholders to mitigate the impacts of tourism on small communities.

Arctic Ocean shipping governance

Shipping includes trans-Arctic passage, Arctic supply shipping, and cruise tourism. Institutions currently governing Arctic shipping are the United Nations (via UNCLOS), the Arctic Council, and the International Marine Organization (IMO) (Brigham, 2013; Buixadé Farré et al., 2014). A major recent development is the adoption by the IMO of the mandatory International Code for Ships Operating in Polar Waters (Polar Code), which began on January 1, 2017 (IMO, 2017).

Governance structures for the Arctic Ocean have been characterized as 'fragmented' (Humbrich, 2013), but 'multidimensional' characterizes it better because of the different actors representing local to international scales. While no single regime exists for Arctic Ocean governance, local and regional stakeholders work with national and international bodies and with private shipping companies to develop management policies (Ng & Song, 2018). Some non-Arctic stakeholders suggest the need for an Arctic treaty, analogous to that which governs Antarctica; others question the desirability of such (Young, 2010, 2016). Arctic governance structures demonstrate the enduring power of states and their respective governments when it comes to decision making on matters related to terrestrial and marine governance.

Knowledge gaps and future priorities for research on governance for Arctic sustainability

The proliferation of governance structures and processes in the Arctic has expanded opportunities for participation and input by northern residents. Given sparse populations in the Circumpolar North, northerners have greater levels of representation compared to citizens in more populated and densely concentrated regions. However, a common complaint is that northerners lack representation centers of power and decision-making. National governments, particularly in the United States and Canada, are located in distant southern capitals, still dominate the political decision-making processes, but this authority is being challenged. In Greenland and Iceland, where national governments are in the north and have jurisdiction over sustainability issues, the situation is different. We must understand where and how higher levels of governance structures support or undermine lower level governance arrangements in the pursuit of sustainability.

The structures, processes, and changes discussed here demonstrate the complexity and dynamism of Arctic governance. From below, decentralization and devolution to sub-national and local governments coupled with the emergence of new Indigenous governments have gradually eroded the authority of national governments. From above, new international and supranational associations pressure national governments to conform to treaties and other agreements. From outside, new governance actors (NGOs, industry associations) contribute to policy development and decision making. Few studies comprehensively address how they collectively impact the Arctic and its inhabitants.

Within this broader context, there are particular governance for sustainability topic that deserve more attention in the Arctic. Efforts to encourage sustainable development depend on coordination between multiple actors within an increasingly complicated policy environment. While expanded governance brings benefits, it also raises questions about the need for multiscalar collaboration. Whether it is sustainable education, health care, resource development, housing, or other policy issues confronting northern communities, creating effective mechanisms for identifying best practices to encourage collaboration must be explored from academic and practical perspectives. The study of best practices is relevant when considering the expansion of horizontal and vertical governance.

Increase in the number of actors demanding a seat at the table demands research that considers the challenges that this presents for achieving sustainable policy outcomes. Many new actors face considerable capacity challenges to fully participate in deliberations. Local and Indigenous communities, for example, often lack the financial means and people to engage regularly in deliberations. Their input, however, is critical because local perspectives and priorities are often different from those of national and supranational actors. While focus has been on the roles that governments play in policy development and implementation, recent trends indicate greater involvement of non-state actors in policy discussions. Their input must be studied in a way that identifies their positive and the negative contributions. Research is needed on how best to incorporate the participation of legitimate actors, including those from beyond the Arctic, in decision making. Special attention should be given to 'equity,' a pillar of sustainability, and to the obstacles to participating in governance for under-represented sectors of society. Gender, age, and ethnicity are critical to how sustainability is experienced and performed, yet research on this dimension of sustainability governance is embryonic.

Collaboration involves governments through formal and informal mechanisms. As such mechanisms grow and evolve, more research will be needed on the dynamics of intergovernmental relations and other governance practices for sustainability. A particularly interesting dynamic is the co-existence of Indigenous and Western worldviews. Little research exists on the relationship between Indigenous and Western perspectives on governance and how collaboration or conflict between these two different approaches could facilitate or impede the sustainable policy development.

It is important to recognize that the Arctic is home to diverse governance models that have different impacts on policy outcomes and on the ability of political actors to pursue sustainability agendas. Most Arctic governance research focuses on individual case studies (either geographical or policy oriented). There is a need for research that systematically compares cases across regions and countries because this will help determine how distinct governance models lead to different outcomes and will identify best governance practices that lead to sustainable policies. Perspectives on what constitutes the governance of sustainability also differ across the Arctic depending on different political cultures and the positions, philosophies, backgrounds, and identities of those involved. It is critical to recognize and explain these differences systematically. Finally, if the goal is to innovate new governance models for the Arctic, which are attentive to lessons learned from the study of best practices, place-based differences, and socio-environmental justice across various axes of difference (gender, age, cultural identity), we must also consider how best to adapt governance to changing socio-SESs.

Notes

1 Searches on Google Scholar were performed on the following terms for years 1992–2016: 'environmental governance,' 'governance for sustainable,' 'sustainable governance,' and 'governance for sustainability'; 'earth systems governance' and 'earth system governance'; 'resilience governance' and 'resilience-based governance'; 'social-ecological governance,' 'socio-ecological governance,' and 'SES governance.' Searches were then performed on these same terms with the addition of the term 'Arctic.' The search was only conducted on publications in English. We recognize that this offers only a proxy of trends in the literature but consider the results to be of interest.

2 The Arctic Five is composed of the five countries that border on the Arctic Ocean (Russia, the United States, Canada, Norway, and Denmark). The other three members of the Arctic Council (Iceland, Finland, and Sweden), the Permanent Participants, and the observer states and organizations are not included in the Arctic Five. The Illulissat Declaration (2017) recognized the impacts of climate change on vulnerable Arctic ecosystems and the livelihoods of local inhabitants and Indigenous communities. It also noted the likely increase of resource development in the wake of climate changes that would make the Arctic Ocean more accessible. The signatories to the Declaration pledged to take steps to protect and preserve the fragile marine environment of the Arctic Ocean by improving the safety of marine navigation, reducing ship-based pollution, and strengthening search and rescue capabilities and capacity through the Arctic.

3 The Aleut International Assocation (AIA), Arctic Athabaskan Council (AAC), Gwich'in Council International (GCI), Inuit Circumpolar Council (ICC), Sami Council, and Russian Association of Indigenous Peoples of the North, Siberia and the Far East (RAIPON).

References cited

Arctic Council. (2017, July 25). *About Us.* http://www.arctic-council.org/index.php/en/about-us/arctic-council/observers

ARR. (2016). *Arctic resilience report* (M. Carson & G. Peterson, Eds.). Retrieved from www.arctic-council.org/arr

Baker, C. (2016). Sustainable governance in a post-secular public sphere: Re-assessing the role of religion as a cosmopolitan policy actor in a diverse and globalized age. *Sustainable Development, 24*, 190–198.

Bennett, M. M. (2015). How China sees the Arctic. Reading between the extraregional and intraregional narratives. *Geopolitics, 20*(3), 645–668.

Beunen, R., Patterson, J., & Van Assche, K. (2017). Governing for resilience: The role of institutional work. *Current Opinion in Environmental Sustainability, 28*, 10–16.

Biermann, F. (2007). Earth system governance as a cross-cutting them of global change research. *Global Environmental Change, 17*, 326–337.

Bjørst, L. R. (2017). Uranium: The road to "economic self-sustainability for Greenland"? Changing uranium positions in Greenlandic politics. In G. Fondahl & G. N. Wilson (Eds.), *Northern sustainabilities. Understanding and addressing change in the circumpolar* (pp. 25–34). Cham, Switzerland: Springer.

Brigham, L. W. (2013), Arctic marine transport driven by natural resource development. *Baltic Rim Economic Quarterly Review, 2*, 13–14.

Brosnan, I. G., Leschine, T. M., & Miles, E. L. (2011). Cooperation or conflict in a changing Arctic? *Ocean Development and International Law, 42*, 173–210.

Brundtland, G. H. (1987). *Our common future. Report of the world commission on environment and development.* Oxford: Oxford University Press.

Buixadé Farré, A., Stephenson, S., Chen, L., Czub, M., Dai, Y., Demchev, D., . . Wighting, J. (2014). Commercial arctic shipping through the Northeast Passage: Routes, resources, governance, technology and infrastructure. *Polar Geography, 37*(4), 298–324.

Chaturvedi, S. (2013). China and India in the 'receding' Arctic: Rhetoric, routes and resources. *Jadavpur Journal of International Relations, 17*(1), 41–68.

Conley, H. A., & Melino, M. (2016). *An Arctic redesign. Recommendations to rejuvenate the Arctic council.* [A Report of the CSIS Europe Program]. Washington, DC: Center for Strategic International Studies.

Dawson, J., Johnston, M., & Stewart, E. (2014). Governance of Arctic expedition cruise ships in a time of rapid environmental and economic change. *Ocean & Coastal Management, 89*, 88–99.

Duyck, S. (2012). Participation of non-state actors in arctic environmental governance. *Nordia Geographical Publications, 40*(4), 99–110.

Duyck, S. (2015). Polar environmental governance and non-state actors. In R. Pincus, S. H. Ali, & S. Gustave (Eds.), *Diplomacy on ice: Energy and the environment in the Arctic and Antarctic* (pp. 13–40). New Haven, CT: Yale University Press.

Exner-Pirot, H., & Plouffe, J. (2013). A proliferation of forums: A second wave of organizational development in the Arctic. *Arctic Yearbook*, 343–345.

Fluharty, D., & Kim, Y. H. (2012). Arctic marine living resources. In O. R. Young (Ed.), *The Arctic in world affairs. A north pacific dialogue on Arctic marine issues.* Seoul and Honolulu: Korea Maritime Institute and East-West Center.

Folke, C., Hahn, T., Olsson, P., & Norberg, J. (2005). Adaptive governance of social-ecological systems. *Annual Review of Environment and Resources, 30*, 441–473.

Fondahl, G., Filippova, V., Savvinova, A., Ivanova, A., Stammler, F., & Hoogensen Gjørv, G. (2019). Niches of agency: Managing state-region relations through law in Russia. *Space and Polity, 23*(1), 49–66.

Fondahl, G., & Wilson, G. N. (2017). *Northern sustainabilities: Understanding and addressing change in the circumpolar world.* Cham, Switzerland: Springer Press.

Forbes, B., & Kofinas, G. (2014). Resource governance. In J. N. Larsne & G. Fondahl (Eds.), *Arctic human development report II. Regional processes and global linkages* (pp. 253–296). Copenhagen: Nordic Council of Ministers.

Gad, U. P., Jakobsen, U., & Strandsberg, J. (2017). Politics of sustainability in the Arctic: A research agenda. In G. F. World & G. N. Wilson (Eds.), *Northern sustainabilities. Understanding and addressing change in the circumpolar world* (pp. 13–23). Cham, Switzerland: Springer.

Garmestani, A. S., & Benson, M. H. (2013). A framework for resilience-based governance of social-ecological systems. *Ecology and Society, 18*(1), 9.

Glasbergen, P. (Ed.). (1998). *Co-operative environmental governance. Public-Private agreements as a policy strategy*. Dordrecht: Kluwer Academic.

Graczyk, P., & Koivurova, T. (2014). A new era in the Arctic Council's external relations? Broader consequences of the Nuuk observer rules for arctic governance. *Polar Record, 50*(254), 225–236.

Griffin, L. (2010). Governance innovation for sustainability: Exploring the tensions and dilemmas. *Environmental Policy and Governance, 20*, 365–369.

Hall, C. M., James, M., & Wilson, S. (2010). Biodiversity, biosecurity, and cruising in the Arctic and sub-Arctic. *Journal of Heritage Tourism, 5*, 351–364.

Hermann, T. M., & Martin, T. (Eds.). (2016). *Indigenous peoples' governance of lands and protected areas in the Arctic*. Cham, Switzerland: Springer.

Hong, N. (2014). Emerging interests of non-arctic countries in the Arctic: A Chinese perspective. *The Polar Journal, 4*(2), 271–286.

Hooghe, L., & Marks, G. (2003). Unraveling the central state but how? Types of multi-level governance. *American Political Science Review, 97*(2), 233–243.

Hsiung, C.W. (2016). China and Arctic energy: Drivers and limitations. *The Polar Journal, 6*(2), 243–258.

Huitric, M., Walker, B., Moberg, F., Österblom, H., Sandin, L., Grandin, U., . . Bodegård, J. (2009). *Biodiversity, ecosystem services and resilience—governance for a future with global changes*. Stockholm: Stockholm Resilience Centre.

Humbrich, C. (2013). Fragmented international governance of arctic offshore oil: Governance challenges and institutional improvement. *Global Environmental Politics, 13*(3), 79–99.

ICC Canada. (n.d.). Retrieved July 28, 2016, from Inuit Circumpolar Council Canada website: www.inuitcircumpolar.com/overview.html

Illulissat Declaration. (2017). Retrieved July 3, 2019, from. https://cil.nus.edu.sg/wp-content/uploads/2017/07/2008-Ilulissat-Declaration.pdf

Indigenous and Northern Affairs Canada. (2017, June 29). Inuit leaders from across Inuit Nunangat and five federal cabinet ministers continued work of Inuit crown partnership committee. *September*. Retrieved from www.canada.ca/en/Indigenous-northern-affairs/news/2017/09/inuit_leaders_fromacrossinuitnunangatand fivefederalcabinetminist.html

International Maritime Organization. (2017). *Shipping in Polar waters*. Retrieved July 28, 2017, from www.imo.org/en/mediacentre/hottopics/polar/pages/default. aspx

Ivanova, A., & Stammler, F. (2017). Mnogoobrazie upravyemosti prirodnymi resursami v Rossiyskoy Arktike [A diversity of ways of governing natural resources in the Russian Arctic). *Sibirskie Istoricheskie Issledovaniya, 4*, 210–225.

Jordan, A. (2008). The governance of sustainable development: Taking stock and looking forward. *Environment and Planning C: Governance and Policy, 26*, 17–33.

Kankaanpää, P., & Young, O. R. (2012). The effectiveness of the Arctic Council. *Polar Research, 31*(17176). Retrieved from dx.doi.org/10.3402/polar.v31i0.17176

Karpouzoglou, T., Dewulf, A., & Clark, J. (2016). Advancing adaptive governance of socio-ecological systems through theoretical multiplicity. *Environmental Science & Policy, 57,* 1–9.

Keil, K., & Knecht, S. (2017). *Governing Arctic change: Global perspectives.* London: Palgrave Macmillan.

Kocho-Schellenberg, J.-E., & Berkes, F. (2015). Tracking the development of co-management: Using network analysis in a case from the Canadian Arctic. *Polar Record, 51*(4), 422–431.

Kofinas, G. P. (2009). Adaptive co-management in social-ecological governance. In F. S. Chapin III, G. P. Kofinas, & C. Folke (Eds.), *Principles of ecosystem stewardship: Resilience-based natural resource management in a changing world* (pp. 77–101). New York, NY: Springer.

Koivurova, T. (2010). Limits and possibilities of the Arctic Council in a rapidly changing scene of Arctic governance. *Polar Record, 46*(237), 146–156.

Koivurova, T. (2011). The status and role of Indigenous peoples in Arctic international governance. In G. Alfresson & T. Koivurova (Eds.), *The yearbook of polar law* (pp. 169–192). Leiden: Martinus Nijhoff.

Koivurova, T. (2013). Multipolar and multilevel governance in the Arctic and Antarctic. *Proceedings of the Annual Meeting of the American Society of International Law, 107,* 443–446.

Kooiman, J., Bavinck, M., Chuenpadgee, R., Mahon, R., & Pullin, R. (2008). Interactive governance and governability: An introduction. *The Journal of Transdisciplinary Environmental Studies, 7*(1), 1–11.

Kuersten, A. (2016). *The Arctic five versus the Arctic council. Arctic yearbook* (L. Heininen, H. Exner-Pirot, & H. J. Plouff, Eds.). Retrieved from www.arcticyear book.com/

Lange, P., Driessen, P. P. J., Sauer, A., Borenmann, B., & Burger, P. (2013). Governing towards sustainability—conceptualizing modes of governance. *Journal of Environmental Policy and Planning, 15*(3), 403–425.

Lawhon, M., & Patel, Z. (2013). Scalar politics and local sustainability: Rethinking governance and justice in an era of political and environmental change. *Environment and Planning C: Governance and Policy,* (31), 1048–1062.

Liu, H. (2018). Influence of the agreement on enhancing international Arctic scientific cooperation on the approach of non-Arctic states to Arctic scientific activities, advances in polar. *Science, 29*(1), 51–60. https://doi.org/doi:10.13679/j.advps.2018.1.00051

Makivik Corporation. (2017, May). Retrieved May 23, 2017, from www.makivik.org/corporate/

Meadowcroft, J., & Bregha, F. (2009). *Governance for sustainable development: Meeting the challenge ahead.* [PRI Project: Sustainable Development]. Retrieved from www.horizons.gc.ca/eng/book/export/html/711

Molenaar, E. (2014). Status and reform of international arctic fisheries law. In E. Tedsen, S. Cavalieri, & R. A. Kraemer (Eds.), *Arctic marine governance. Opportunities for transatlantic cooperation.* Heidelberg: Springer.

Newig, J., & Fritsch, O. (2009). Environmental governance: Participatory, multi-level—and effective? *Environmental Policy and Governance, 19*, 197–214.

Ng, A. K. Y., & Song, D.-W. (2018). Special issue on 'Arctic shipping, transportation and regional development. *Maritime Policy & Management, 45*(4).

Pashkevich, A., Dawson, J. S., & E.J. (2015). Governance of expedition cruise ship tourism in the Arctic: A comparison of the Canadian and Russian Arctic, *Tourism in Marine Environments*, 10, 225–240.

Paterson, M., Humphreys, D., & Pettiford, L. (2003). Conceptualizing environmental governance: From interstate regimes to counter-hegemonic struggles. *Global Environmental Politics, 3*(2), 1–10.

Peck, J., & Tickell, A. (2002). Neo-liberalizing Space. *Antipode, 34,* 380–404.

Petrov, A. N., BurnSilver, S., Chapin, F. S. III, Fondahl, G., Graybill, J. K., Keil, K., .. Schweitzer, P. (2016). Arctic sustainability research: Toward a new agenda. *Polar Geography, 39*(3), 165–178.

Plummer, R., Armitage, D., & de Loë, R. (2013). Adaptive comanagement and its relationship to environmental governance. *Ecology and Society, 18*(1), 10–5751.

Plummer, R., & Baird, J. (2013). Adaptive co-management for climate change adaptation: Considerations from the Barents region. *Sustainability, 5*(2), 629–642.

Poelzer, G., & Wilson, G. N. (2014). Governance in the Arctic: Political systems and geopolitics. In J. N. Larsen & G. Fondahl (Eds.), *Arctic human development report: Regional processes and global linkages*. Copenhagen: Norden.

Saami Council. (2017, July 28). Saami Council (SÁMIRÁÐÐÁI) website.

Schatz, V. J., Proelss, A., & Liu, N. (2019). The 2018 agreement to prevent unregulated high seas fisheries in the central Arctic Ocean: A critical analysis. *The International Journal of Marine and Coastal Law, 34*(2).

Shadian, J. (2014). *The politics of Arctic sovereignty: Oil, ice and Inuit governance*. Oxon: Routledge.

Shibata, A., & Raita, M. (2016). An agreement on enhancing international arctic scientific cooperation: Only for the eight arctic states and their scientists? *Yearbook of Polar Law, VII*, 129–162.

Shiroyama, H., Yarime, M., Matsuo, M., Shroeder, H., Scholz, R., & Ulrich, A. (2012). *Governance for Sustainability: Knowledge Integration and Multi-Actor Dimensions, 7*, 45–55.

Smith, J. (2014). Intuitively neoliberal? Towards a critical understanding of resilience governance. *European Journal of International Relations, 2*, 402–426.

Smith, P. (2013). Natural resource comanagement with aboriginal peoples in Canada: Coexistence or assimilation. In D. Durrant & G. Johnson (Eds.), *Aboriginal peoples and forest lands in Canada* (pp. 89–113). Vancouver: UBC Press.

Solli, P. E., Wilson Rowe, E., & Lindgren, W. Y. (2013). Coming into the cold: Asia's arctic interests. *Polar Geography, 36*(4), 253–270.

Stonehouse, B., & Snyder, J. (2010). *Polar tourism: An environmental perspective*. Bristol: Channel View Publications.

Stossel, S., Tedsen, E., Cavalieri, S., & Riedel, A. (2014). *Environmental governance in the maritime Arctic, in Arctic marine governance: Opportunities for transatlantic cooperation* (E. Tedsen, S. Cavalieri, & R. A. Kraemer, Eds.). Heidelberg: Springer.

Tasker, J. P. (2017, July). Supreme court quashes seismic testing in Nunavut but gives green light to Enbridge pipeline. *CBC News*. Retrieved from www.cbc.ca/news/politics/supreme-court-ruling-Indigenous-rights-1.4221698

Tesar, C., Dubois, M. A., Sommerkorn, M., & Shetakov, A. (2016). Warming to the subject: The Arctic Council and climate change. *The Polar Journal, 62*(2), 417–429.

UNEP (United Nations Environment Program). (1992). Rio declaration on environment and development. *UN Doc, 151*(26).

Veld, R. J. (2013). Transgovernance: The quest for governance of sustainable development. In L. Meuleman (Ed.), *Transgovernance. Advancing sustainability governance*. Dordrecht: Springer.

Voss, J. P., & Borenmann, B. (2011). The politics of reflexive governance: Challenges for designing adaptive management and transition management. *Ecology and Society, 16*(2).

Watt-Cloutier, S. (2015). *The right to be cold: One woman's story of protecting her culture, the Arctic and the whole planet*. Toronto: Allen Lane.

Wehrmann, D. (2016). The polar regions as "barometers" in the Anthropocene: Towards a new significance of non-state actors in international cooperation. *The Polar Journal, 6*(2), 379–397.

White, G. (2006). Cultures in collision: Traditional knowledge and Euro-Canadian governance processes in Northern land-claim boards. *Arctic, 59*(4), 401–414.

Williams, C. C., & Millington, A. C. (2004). The diverse and contested meanings of sustainable development. *The Geographical Journal, 170*, 99–104.

Wilson, G. N. (2008). Nested federalism in arctic Quebec: A comparative perspective. *Canadian Journal of Political Science, 41*(1), 71–92.

Wilson, G. N., & Alcantara, C. (2012). Mixing politics and business in the Canadian arctic: Inuit corporate governance in Nunavik and the Inuvialuit settlement region. *Canadian Journal of Political Science, 45*(4), 781–804.

Wilson, G. N., & Selle, P. (2019). *Indigenous self-determination in Northern Canada and Norway. IRPP Study: Canada's changing federal community* (No. 69, pp. 1–38). Retrieved from Institute for Research on Public Policy website: https://irp.org/research-studies/Indigenous-self-determination-in-northern-canada-and-norway/

Young, N., & Matthews, R. (2007). Resource economies and neo-liberal experimentation: The reform of industry and community in rural British Columbia. *Area, 39*(3), 176–185.

Young, O. R. (2010). Arctic governance—Pathways to the future. *Arctic Review on Law and Politics, 2*, 164–185.

Young, O. R. (2013). The evolution of Arctic Ocean governance: Challenges and opportunities. In J. D. K. & Y. H. Kim (Eds.), *The Arctic in world affairs. A north pacific dialogue on the future of the Arctic* (pp. 267–298). Seoul and Honolulu: Korea Maritime Institute and East-West Center.

Young, O. R. (2016). The shifting landscape of arctic politics: Implications for international cooperation. *The Polar Journal, 6*(2), 209–223.

Young, O. R., & Kim, Y. H. (2012). Informal arctic governance mechanisms: Listening to the voices of non-arctic ocean governance. In O. R. Young (Ed.), *The Arctic in world affairs. A north pacific dialogue on Arctic marine issues* (pp. 275–303). Seoul and Honolulu: Korea Maritime Institute and East-West Center.

Zeijl-Rozema, A., Cövers, R., Kemp, R., & Martens, P. (2008). Governance for sustainable development: A framework. *Sustainable Development, 16*, 410–428.

6 Methodological challenges and innovations in Arctic community sustainability research

Gary Kofinas, Shauna BurnSilver, and Andrey N. Petrov

Methods broadly defined

Despite advances in the practice of Arctic sustainability science, significant challenges remain. Here, we focus on those associated with community sustainability research in the Arctic. We identify innovative strategies for transcending problems in ways that contribute to collaboration, knowledge production, and social learning. We examine five arenas: making research relevant to communities, co-producing knowledge, accounting for scale and linkages, developing indicators, and realizing praxis.

Research on Arctic community sustainability is similar to and different from conventional disciplinary research. Maintaining high standards of rigor is important when collecting and analyzing qualitative and quantitative data to support conclusions. A sustainability focus demands holistic approaches to frame and analyze problems that link social, economic, and environmental domains. It is expected that findings will be policy and community relevant and delivered to inform decision making. Working with communities to understand and chart paths toward sustainability involves additional challenges, such as collaboration with local peoples, sensitivity to local culture, and guarding against unintended consequences that may harm communities, their residents, or their environments. These challenges commonly require asking what constitutes rigorous research and legitimate knowledge (Berkes & Berkes, 2009; Tengö, Brondizio, Elmqvist, Malmer, & Spierenburg, 2014).

We define research methods broadly. Whereas research methodology typically describes data collection and data analysis, we include aspects of collaborative engagement at multiple research stages and how results are communicated and utilized. The five challenges addressed in this chapter are not inclusive, but they capture important aspects of community sustainability research requiring greater attention. While this examination of methodological challenges in community sustainability research is aimed at research professionals, others directly and indirectly involved in research (funders and community leaders) may also benefit.

The need

The need for a better understanding of the determinants of successful community sustainability research methods follows from unprecedented environmental and social changes in the Arctic (Kumpula, Pajunen, Kaarlejärvi, Forbes, & Stammler, 2011; Moon et al., 2019). Historically, Arctic research has not been attentive to local communities' needs or priorities and instead sought to advance basic science and generalizable knowledge. In the past decade, a newer research paradigm considers how researchers can address the real-world needs of communities. As noted by sustainability scholars, the dichotomy is false: basic and applied science need not be mutually exclusive but can be complementary (i.e., "Pasture's quadrant," Figure 6.1) and ideally should iteratively inform each other (Stokes, 1997; Clark, 2007).

Despite recent revelations and actual shifts in practice, many researchers are unaware of the embedded challenges of executing community sustainability research. Early scholarly literature idealized the potential benefits of community sustainability studies without fully addressing the significant barriers (Brunet, Hickey, & Humphries, 2014). These challenges are multi-dimensional and interrelated and include identifying an appropriate research problem and research team, allocating research funding; learning from diverse, disparate knowledge systems; and addressing unequal power relationships between communities and researchers. Also important is the apparent disinterest of some local decision makers in science and the disinterest of some scientists in local ways of knowing. There is, however, reason for hope as academic and community practitioners are currently gaining experience and experimenting with innovative Arctic community sustainability research methods.

Methods for community sustainability research writ large

Sustainability is often conceptualized as a three legged-stool with economic, social, and ecological components, a starting point that forces researchers to select and devise methods that account for present and possible future conditions of each component and their interactions. The rationale for holism in sustainability research aims to overcome solutions that address only one domain, which may result in unanticipated outcomes in others (Chapin, Stuart, Kofinas, & Folke, 2009). In short, robust solutions require comprehensive consideration of several perspectives and the insights of multiple disciplines (Frodeman, 2011).

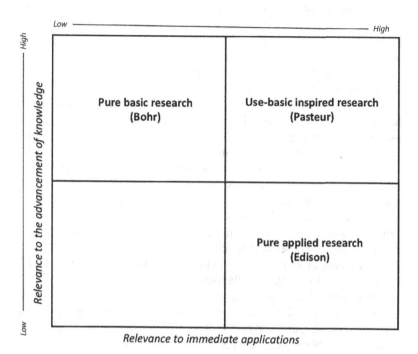

Figure 6.1 Pasteur's quadrant. Louis Pasteur undertook basic science to understand infectious disease while also exploring the evolution of vaccination. His work is in contrast to that of Niels Bohr, the Danish physicist who studied quantum theory and atomic structure, and Thomas Edison, the inventor of electric light bulb. Hence, Pasteur's quadrant represents a goal of sustainability science—to conduct basic science while considering the application of findings to achieving sustainability. (After Stokes, 1997; Clark, 2007.)

Working with multiple disciplines raises questions about knowledge integration. Distinctions among multidisciplinary, interdisciplinary, and transdisciplinary research are highlighted in the literature, with some arguing that transdisciplinary approaches are preferred in sustainability science (Lang et al., 2012), especially when problem-oriented analytical frameworks are valued instead of cobbled-together, discipline-based approaches with different underlying assumptions and biases (i.e., stapled interdisciplinarity) (Ledford, 2015). Whereas interdisciplinary research integrates disciplines into a coherent research cluster that provides a new framework for understanding, transdisciplinarity includes multiple forms of knowledge, including those outside conventional scientific traditions (Holm et al.,

2013). Transdisciplinarity may not imply agreement or synthesis across disciplinary visions, instead requiring collaborative processes that co-produce knowledge and understandings of what sustainability means and how to achieve it (Klein, 2004). Achievable in some cases, the dynamics related to interacting worldviews and power can be a significant barrier to success in others (Vlasova & Volkov, 2016).

A transdisciplinary social-ecological systems (SESs) approach in sustainability science increasingly applies the concept of resilience, which examines endogenous and exogenous forces for change, slow and fast variables, regime shifts, stabilizing and amplifying feedbacks, and critical interactions of system components. Thus, more organic understandings of complex dynamic interactions, non-linear responses, and processes of emergence replace mechanistic representations of systems with independent and dependent variables (Holling, 2001). In this approach, SESs are self-organized, complex, and adaptive systems that adjust their fundamental properties and responses to change (Levin, 1999).

The concept of complexity recognizes the "wicked" nature of social-ecological problems (Rittel & Webber, 1973; DeFries & Nagendra, 2017). Wickedness entails the prevalence of non-linear changes, uncertainty of drivers and outcomes, and inability to truly "solve" a problem (Zijp, Posthuma, Wintersen, Devilee, & Swartjes, 2016) because partial solutions to problems create new problems or frame the problem anew. Consequently, complex problems can never be truly solved. The wickedness of complex problem requires focusing on building resilience to uncertain future conditions or decreasing vulnerability to current conditions. The recent framing of sustainability within resilience thinking has highlighted the potential for crossing critical thresholds, with resultant regimes shifts, and the need for transformative change (Walker & Salt, 2006, Kofinas et al., 2013). Resilience questions sustainability, emphasizing that systems move through stable states (Walker & Salt, 2006) and suggesting that societies must identify which aspects of their SESs are desirable and should be sustained and which should be transformed. A recent concept, resilience pathways, suggests how planning for change and loss brings opportunities through reinvention (Robinson & Carson, 2016). Critiques of resilience suggest that SES perspectives understate power and equity dynamics, ignore history, and burden marginalized people (Chandler, 2013; Reid, 2013; Olsson, Jerneck, Thoren, Persson, & O'Byrne, 2015). Critics ask who defines sustainability and how equitable engagement in sustainability research may be achieved (Gad & Strandsbjerg, 2018).

Normative, or value-based, approaches are part of sustainability science (Cole, 1999). Wiek, Farioli, Fukushi, and Yarime (2012) noted how research can focus on whether particular actions are sustainable or can be oriented toward generating actions for sustainability. Both approaches highlight outcomes instead of focusing on process-oriented approaches to sustainability research (e.g., what processes are conducive to co-learning between local knowledge holders, researchers, and policymakers?). Lang et al. (2012) emphasized that new methods of producing knowledge and making decisions are critical for sustainability science. Like others (Petrov et al., 2017; BurnSilver, Magdanz, Stotts, Berman, & Kofinas, 2016; van der Hel, 2016), they argue that non-academic actors must be engaged in research to draw on best available knowledge, resolve differences in values and preferences and create ownership of problems and solutions (Lang et al., 2012). In short, stakeholder partnerships, not just stakeholder involvement, are necessary.

The Arctic context for community sustainability research

Methodological challenges of Arctic community sustainability research derive from the demands of sustainability science and from the region's unique social, ecological, economic, and political contexts (Larsen & Fondahl, 2015; ARR, 2016; AMAP, 2017a, 2017b, 2018).

For northern Indigenous peoples, colonization by southern-based governments created mistrust in Western/Southern institutions, including justice, education, medicine, and "Western science." Strong cultural identities, ongoing efforts at self-determination, and emerging self-confidence in the value of Indigenous knowledge, mean that community residents increasingly create community-relevant research agendas. Land claim settlements and self-governance agreements (e.g., Alaska's Native Claims Settlement Act, Canadian comprehensive land claims agreements, and Greenland's self-rule) and co-management arrangements (e.g., the Alaska Eskimo Whaling Commission) have formalized Indigenous rights. Many social science professional associations have research ethics statements (IASSA, 2019), and funding organizations articulate ethical principles in Arctic research (IASSA, 2019). In some regions, researchers seeking approval of a project find that community consent to conduct research is legally formalized, such as through the Explorer's Act of Canada. However, these requirements are inconsistent across the Arctic.

Community residents are aware of how colonial approaches played out through research have harmed human and ecosystem health (e.g., experiments in human radiation exposure, relocation of entire communities for wildlife conservation, residential schools, and proposed nuclear testing). Researchers hoping to engage communities today may find local leaders who understand research legacies, the difference between well and poorly planned studies, who question the legitimacy of traditional ecological knowledge by past investigators and desiring follow-up communication.

Within communities, common challenges are burnout and fatigue from what is perceived as too many studies, too many researchers asking questions, and researchers who steal information from communities for their own benefit. And, while climate change research is important, community residents perceive the sidelining of other, more pressing issues, such as health care or economic development (Bali & Kofinas, 2014). Adding confusion are public calls for more central roles by communities in research. An apparent paradox has emerged: local leaders ask for communities to be involved in study design and implementation, but residents sometimes demonstrate limited participation. We suggest that these seemingly inconsistent forces result from the limited authority of communities in the research process and the choice of research methods employed.

Five challenges and areas of innovation

The conditions described earlier point to five methodological challenges of Arctic community sustainability research, which highlight how methods can ultimately determine whether a particular study provides meaningful collaboration, insights, and learning opportunities for scientists, residents, and policymakers (Figure 6.2).

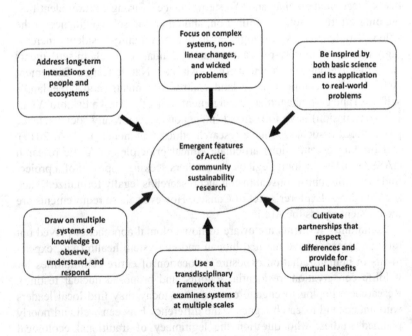

Figure 6.2 Elements of Arctic community sustainability research

Making research relevant

Relevance refers to making the objectives, questions, and methods of research meaningful to communities, policymakers, and researchers. Being relevant—to local concerns, to information needs, to areas of interest—increases engagement among everyone in the research process (Eamer, 2004). Given the diversity of players in a community sustainability research enterprise, however, finding common relevance can be difficult.

Making research relevant to communities and policy requires transformations in sustainability research practice. First, Arctic research culture must change to develop studies valuable to local residents. This requires researchers to develop long-term relationships with communities, share common experiences, and achieve mutual respect. While such relationships may take years to cultivate, "seed" and "network" grants can help achieve understanding and trust.

Strong local leadership helps communities to make research relevant. One strategy is to develop structured processes with skilled facilitators (Vargas-Moreno, Fradkin, Emperador, & Lee, 2016). Another is to formalize community–researcher partnerships, such as the Community Partnerships for Self-Reliance Program in Alaska (Chapin III, Corrine, Brinkman, Bronen, & Cochran, 2016). This approach gives communities greater control of research agendas and methods selection. A third strategy is for communities to self-direct research projects, determining the scope and methods and selecting researchers. However, large-scale funding agencies have not typically accommodated these arrangements. Increasingly, communities and regional organizations receive funding, manage their own projects, and achieve excellent results.

Co-producing knowledge

Knowledge co-production is the collaborative process of combining perspectives, data, and theories of different knowledge systems to generate a better understanding of a problem or phenomena (Kofinas et al., 2002; Armitage, Berkes, Dale, Kocho-Schellenberg, & Patton, 2011; Euskirchen et al., in Press). Co-production typically includes local knowledge holders, academics, agency scientists, and policymakers to understand system dynamics, response options, and the implications of solutions to problems. Van der Hel (van der Hel, 2016) describes three distinct rationales for practicing coproduction: (1) to enhance scientific accountability to society, (2) to ensure the implementation of scientific knowledge, and (3) to include the knowledge, perspectives and experiences of "extra-scientific actors" (e.g., non-scientists) in scientific knowledge production. He notes that the heterogeneous conception of knowledge co-production provides helpful ambiguity. Knowledge co-production can also be pivotal for decolonizing research and addressing inequities among knowledge holders (Petrov et al., 2017).

Pohl et al. (2010) identified three key challenges facing knowledge co-production with non-academics: addressing asymmetric power relations, the process of integrating or interrelating different perspectives, and promoting negotiated orientation toward sustainable development. Nel et al. (2016) discussed creating permeable knowledge boundaries to satisfy the needs of multiple groups while ensuring the integrity of contributing knowledge. Reid, Berkes, Wilbanks, and Capistrano (2006) emphasized humility by researchers as key to developing trust, with humility shown in multiple researcher interactions with community members, such as the duration of stay, communicating mode (listen, don't just talk), and transportation mode (be willing to get out and walk). Regarding governance, Armitage et al. (2011) explored co-management institutions in knowledge co-production, noting that institution building creates a decision-making arena that enables knowledge co-production.

Expectations about knowledge integration is defines complications with knowledge co-production if integration is defined as achieving a singular understanding. The idea of "melding" knowledge systems suggested benefits from integrating traditional ecological knowledge with science (Freeman & Carbyn, 1988), but integration has been questioned (Nadasdy, 1999; Voorberg, Bekkers, & Tummers, 2015) because dominant players can draw selectively without fully accounting for different cultural views and assumptions.

Perhaps integration should be redefined, for example, through parallel processes of knowledge co-production, whereby knowledge holders with differing perspectives work to inform, enlighten, and inspire (Voorberg et al., 2015). This approach is especially useful if those with differing perspectives strive to identify new research questions, alternative hypotheses, unrecognized assumptions, data gaps, and unexplored relationships.

An effective strategy for managing community-researcher co-production is to create project steering committees of well-informed and respected local residents and researchers who can co-navigate questions related to project goals, logistics, and outcomes, leading to trust and investment in the collaboration (Kofinas et al., 2016a). Steering committees may also overcome engagement with only highly visible, formal leaders who may be overburdened or have competing demands. Knowledge co-production is best undertaken with the assumption that the overarching objective is mutual learning through effective communication.

Accounting for scale and linkages

Arctic community sustainability should also account for history and relationships within and across multiple geographic scales and social levels

(Neumann, 2009). Geographic scale refers to the extent or size of the system, such as an ecosystem, a watershed, a community's homelands, or the entire Earth (Rotmans & Rotmans, 2003). Social levels, in contrast, convey the hierarchy of social organization, such as a household, neighborhood, community, city, state or province, national, or supra-national entities. Social forces at specific levels may unfold as international-level geopolitical processes, national-level policies, and regional boom–bust cycles, affecting communities through downscaled interactions. At sub-community levels, complex sets of entities and relationships, such as households, kin relationships, whaling crews, work teams, and individual fishers and hunters, interact to shape community-level perceptions, knowledge, values, preferences, and actions. Processes or structures relevant to geographic scale or social level may be hierarchical or lateral. In governance, lateral and polycentric interactions aid in understanding how communities interact with other communities, non-governmental organizations, and boundary organizations. Institutions may act at different spatial scales or social levels to facilitate adaptation and political action.

Temporal dynamics apply to scale and level. The role of path dependence (i.e., how prior decisions dictate present and future choices; North, 1990) in shaping sustainability outcomes is the reason for historians to be integral to the research process (Mahoney, 2000). There is also a need to account for ways community action affects phenomena at larger scales, such as how Indigenous community advocacy for conservation and habitat protection has influenced national-level public opinion (e.g., Gwich'in efforts to stop oil development in Alaska's Arctic Wildlife Refuge).

SOCIAL NETWORK ANALYSIS

Social networks represent relationships and consist of sets of nodes and ties (Borgatti, Mehra, Brass, & Labianca, 2009). Nodes represent an entity of interest at some social level, such as an individual, a household, or a stakeholder. Ties represent connections between entities. Ties in the Arctic community context might include flows of information, financial support, or the sharing of wild foods. Ties may be described according to simple presence or absence, their magnitude (frequency, strength), directionality (reciprocal or one directional), or duration (change over time). The basic assumptions behind social network analysis are threefold. First, social relationships represent social capital, which matters in the context of change (Lin, 2002; Putnam, 2007) and may be important in sustainability outcomes (Armitage, 2005; Adger, 2010). Second, structures and characteristics of social networks may act as drivers of other social or economic outcomes (Bodin & Crona, 2009). Finally, the reverse may also be true: interactions between other system attributes (social, economic, or ecological) may result in network characteristics relevant to sustainability. For example, networks of social and economic activity within Arctic mixed economies can highlight dependence on high-producing super-hunters, or so-called "super-households" (Thornton, 2001; Wolfe et al., 2009). These assumptions create the foundation for social network methodology used in the Arctic, commonly framed as vulnerability and resilience research and oriented toward understanding social and ecological linkages.

While numerous Arctic datasets relate the dynamics of subsistence harvest and productivity (Kruse, 1991; Thornton, 2001; Wheeler & Thornton, 2005; Wolfe et al., 2009), this research has remained separate methodologically from meaningful community narratives about livelihoods, identity, human–animal relationships, and social relationships of sharing and cooperation. However, social network analysis is increasingly used to highlight resource and policy linkages between and within communities. Early work by Wenzel (1995) and Magdanz, Utermohle, and Wolfe (2002) situated subsistence hunting and fishing in larger processes of sharing, exchange, and distribution. Recent social network-based research highlights how many social relationships translate directly into community resource access (Ziker & Schnegg, 2005; Collings, 2011; Natcher, 2015; BurnSilver et al., 2016), adaptive capacity and sensitivity (BurnSilver & Magdanz, 2019), health (Dombrowski, Channell, Khan, Moses, & Misshula, 2013), and equity outcomes (Ready & Power, 2018). Kofinas et al. (2016b) illustrated that household networks scale up to link multiple Alaskan communities together through flows of wild foods.

Building governance capacity is important in sustainability science. Social network theory provides structure for thinking about how connectivity among decision-making entities within communities, among communities, and with larger scale social levels may define governance outcomes. Social capital and network theorists suggest that communities need horizontal (within one social level) and vertical (between social levels) linkages to gain access to institutional resources (Young, 2002) and decision makers with the power to affect decisions that improve, or impinge on, life within Arctic communities. While much research on Arctic governance (Keskitalo & Kulyasova, 2009; Young, 2012) and co-management (Berkes & Armitage, 2010; Meek, 2013) uses network language, applications of network methods to analyze governance patterns are still rare, although this kind of research in non-Arctic contexts is rapidly expanding (Bodin & Crona, 2009; Lubell, Robins, & Wang, 2014).

These trajectories have significant potential for Arctic community sustainability research. Social network theory links people across geographic scales and social levels and incorporates discussions of power relationships that affect stakeholder network performance. Similarly, a network lens can embed people in Arctic places, connecting them to resources and other people, strengthening cultural narratives, and illustrating potential vulnerabilities. However, the collection of social network data, particularly at the community level, is labor intensive and requires large sample sizes because partial data can miss aspects of a social network and misrepresent the system. Consequently, a respondent's willingness to contribute data is critical, as is the trust between respondents and researchers.

PANARCHY

Social and biogeophysical systems are commonly characterized by top-down and bottom-up controls (Allen, Angeler, Garmestani, Gunderson, & Holling, 2014; Allen et al., 2016; Berkes & Ross, 2016). Gunderson and Holling (2002) provide a variant on hierarchy theory by describing multi-scale socio-ecological interactions with panarchy, which identifies patterns of stability and change within and across scales as cyclical processes of conservation, release, exploitation, and reorganization. Panarchy (Figure 6.3)

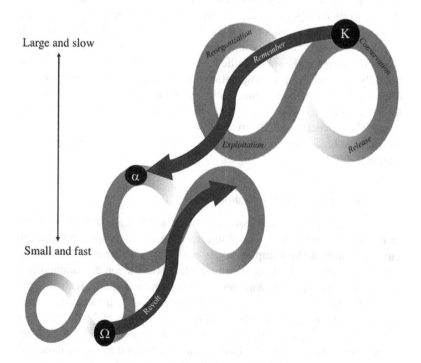

Figure 6.3 Panarchy of nested subsystems at different stages of their adaptive cycles and operating at different scales with points of interaction (after Gunderson & Holling, 2002; Chapin III, Kofinas, & Folke, 2009)

differs from top-down control by accounting for adaptive cycles concurrent with and interacting across levels (Garmestani & Cabezas, 2008). Panarchy explains nested systems, conditions that may result in bottom-up change, and the surprise and uncertainty of such systems.

Methods for studying cross-scale interactions using the panarchy heuristic are descriptive and based on case studies. For example, Moen and Keskitalo (2010) explored multiscalar interlocking panarchies related to boreal forest SES in Sweden. Berkes and Ross (2013) included psychological considerations for examining Arctic Indigenous resilience as part of panarchy. Alessa, Kliskey, and Williams (2010) used an empirically based approach to consider community sustainability and panarchy regarding freshwater and human wellbeing.

Descriptive case study approaches have limitations, particularly for theory building. Simulation modeling offers an alternative method for exploring multi-scale interactions with more formal analysis (Scheffer et al., 2018) and agent-based models (ABMs) offers less mechanistic community representation for capturing human choice and non-linearity. Balbi and Giupponi (2010) argued that ABMs have potential to link social and environmental models while incorporating the effects of micro-level decision making and examining collective responses to policies. Only a few studies have concretely (and collaboratively with communities) applied ABM methodology (Berman, Nicolson, Kofinas, Tetlichi, & Martin, 2004), and its potential is not fully realized. Formal computation models, however, have their own set of problems as a method in community sustainability studies because of the limited availability of long-term data (Allen et al., 2014). Technical aspects of model construction can divert attention away from a focus on sustainability, and the "black-box" problem also exists (Nicholson, Starfield, Kruse, & Kofinas, 2002).

Developing sustainability indicators

Indicators characterize the current state of a system and track changes in its characteristics over time. Indicators are useful in their potential to direct research, make complex issues more easily understood by society, and inform decision making (Petrov et al., 2016; Carson & Sommerkorn, 2017). Sustainability indicators represent systems in holistic, realistic, participatory, and systemic ways (Bell & Morse, 2001); incorporate social, economic, environmental, and institutional elements; and may measure sustainable development progress (Valentin & Spangenberg, 2000). Useful indicators are measurable, based on accessible and available data, and interpretable (Larsen, Schweitzer, & Fondahl, 2010). Constructing indicators is challenging because of the complex nature of social, economic, and environmental systems; their dissimilarities among places and communities; and the processes they describe. Arctic sustainability indicators should also account for communities' adaptive capacities, crucial for indices or compound indicators in which different elements must be weighted and combined (Hamilton & Lammers, 2011; Petrov et al., 2017).

Sustainability indicators were focal in prior research (Wu & Wu, 2012) and introduction of the United Nations' Millennium Development Goals (MGDs) and the Sustainable Development Goals (SDGs) have improved how their categorization and measurement (UN, 2007; UN, 2015). Development of global, regional, and local indicators has produced much literature (Sachs, 2012; Fehling, Nelson, & Venkatapuram, 2013; Griggs et al., 2013). Most indicators are applied at the level of populations and data are drawn from national- or regional-level sources and include the Environmental Sustainability Index, Living Planet Index, and the Genuine Progress Indicator. The UN Agenda 2030 documents 232 official indicators, covering all 17 Sustainability Goals (UN, 2015).

Although several Arctic countries have introduced the UN's SDGs indicator frameworks (Weitz, Persson, Nilsson, & Tenggren, 2015; Canada, 2008; Halonen et al., 2017), they have not been re-tailored for the Arctic. There is no Arctic-specific sustainability index that incorporates social, economic, and ecological components to track Arctic sustainable development (Ozkan & Schott, 2013; Petrov et al., 2017). While indicators of Arctic geophysical systems, biodiversity, ecosystems, and human wellbeing are monitored, only some have developed integrated indicators to address Arctic SESs, and fewer have developed community-level indicators.

The second Arctic Human Development Report (Larsen & Fondahl, 2015) proposed a framework for pan-Arctic monitoring with Arctic specific. As a result, the Arctic Social Indicators project characterizes and tracks human development (Larsen et al., 2010; Larsen, Schweitzer, & Petrov, 2015) in six domains of human wellbeing.

Only a few studies explicitly measure social-ecological sustainability in the Arctic. Early research in Alaska developed and applied indicators defined by community goals (Kruse et al., 2004). Recent research found that those same sustainability goals are viewed by Alaska North Slope community leaders as consistent with current goals (Blair, 2017). Votrin (2006) designed 30 sustainability indicators for the Russian Arctic, but this work remains unknown to a broader Arctic audience. A promising initiative to develop an Arctic Urban Sustainability Index (AUSI) is being pursued (Suter, Schaffner, Giddings, Orttung, & Streletskiy, 2017) that capitalizes on emerging Arctic urban sustainability literature (Orttung & Laruelle, 2017) to create comprehensive sustainability monitoring systems for Arctic cities and towns.

If indicator criteria are developed collaboratively with communities, the process may yield robust sustainability metrics (Carson & Sommerkorn, 2017). But given the Arctic's cultural and geographic diversity, selected indicators may not transfer well. Similarly, with an increased cash economy, livelihood strategies, life goals, capacities, and social-ecological outcomes diversify within communities and households alongside subsistence (AHDR, 2015). Indicators based on small samples or mean values from census data cannot represent the heterogeneity of activities and sustainability outcomes within communities (BurnSilver & Magdanz, 2019).

The examples suggest that progress towards developing and applying Arctic community sustainability indicators exists, but major challenges persist, including data availability, data quality, interoperability of data and definitions, and community engagement in indicator development and use (Petrov et al., 2017). Differences between biogeophysical and social and economic datasets and the diversity of methods used to collect and report indicators makes comparability difficult, further underscoring the need for collaborative, interdisplinary, and transdisciplinary efforts.

Focusing on praxis

Praxis is the process by which theory, including research findings, becomes part of decision making (Lather, 1986). The first element relates to the dominant paradigm of knowledge production and the culture of policymaking. Conventional science and policymaking are rigid, and paradigm shifts and culture change are slow. The expectation that science be policy-neutral exacerbates the problem (Neff, 2009), as does the strained trust relationships between Indigenous peoples and scientists. Despite these issues, Arctic researchers and communities are shifting the nature of these relationships (Huntington, 2011).

Making changes includes modifying existing organizational reward systems (i.e., how funding is awarded and promotion of individuals occurs) to attend to community needs. This change is needed in higher education, especially in the curricula of many institutions, where sustainability science is still taught with discipline-based research methods that provide limited teams and community research training. Re-orienting academic and agency science to generate products and develop processes that engage stakeholders to inform policymaking is crucial.

The second aspect involves formal institutional arrangements, which define people's roles, shape behavior, and convey power and authority in decision making (Ostrom, 2005; Young, 2016). Many Arctic institutions have redefined the rights of Indigenous peoples through land claims, self-government agreements, and co-management. In some cases, these institutions have fostered stronger relationships among scientists, community leaders, and local residents, creating space for knowledge exchange (Armitage et al., 2011). Formal institutions have also given greater legitimacy to Indigenous knowledge, informing the policy process.

The third aspect relates to the development and use of tools and strategies that facilitate long-term thinking and social learning for sustainability decision making. Advances include community-based monitoring programs that document local and Indigenous knowledge of observations and understandings of change using cutting-edge technologies (Brubaker, Berner, & Tcheripanoff, 2013). However, community-based monitoring is rarely linked to sustainability-relevant indicator programs and policy processes. Participatory scenario planning, now widely used across the Arctic (Flynn, Ford, Pearce, & Harper, 2018; Nilsson et al., 2019), illustrates future conditions and the consequences of tradeoffs associated with various policy alternatives (Kruse & White, 1995). When a scenario planning process is well executed, it may aid stakeholders with different expertise to develop joint human–ecosystem management strategies and adaptation options (Nilsson et al., 2017; Planque et al., 2019), identify research needs (Vargas-Moreno et al., 2016), and apply the findings of climate change downscaling models (Ernst & Riemsdijk, 2013). Flynn et al.'s (2018) review of Arctic scenario analysis found that integrating different knowledge systems and attention to cultural factors positively affects a program's utility and acceptance by stakeholders. The long-term utility of these methods in supporting development of change-resilient pathways will depend on the quality of future projections, decision-support systems, their attention to cultural factors, and refinement of methods that facilitate participant dialogue.

Conclusion

Methods for Arctic community sustainability research are developing alongside sustainability science. Methods continue to evolve—from focus on the human impacts of climate change to community-based and community-driven transdisciplinary approaches with direct links to the policy process. The complexities underpinning sustainability science, such as post-normality, wicked problems, or panarchy, drive methodological diversity. The main unit of analysis for sustainability science, the SES, underscores the holistic nature of sustainability research and its close attention to system dynamics embedded in the concepts of resilience and adaptation. Understanding SES dynamics necessitates "making sense together" (Klein, 2004) and demands methods that engage non-academic actors and forge community partnerships. This includes, most importantly, Indigenous stakeholders, rights holders, and knowledge holders working with Western researchers to combine ways of knowing through a co-production process.

While Arctic sustainability research methods are unique, because of the nature of Arctic SESs, they are highly relevant globally. Research practices of community partnership and knowledge co-production are becoming well-established worldwide. There are many challenges that will define the future of methodological innovation in Arctic sustainability science, including relevance, knowledge co-production, scale and linkages, indicators, and praxis. These innovations will address missing links and gaps in sustainability while fostering important evolution toward polycentrism, decolonization, transdisciplinarity, community relevance, and practical applicability. Scale and linkages within and between SESs are crucial, but still relatively underexplored, components of sustainability knowledge. Indicators of sustainability, both a process and an outcome, are in high demand (Vlasova & Volkov, 2016) and will be helpful tools to shape decision making and community development pathways, including working to advance the UN SDGs. Notable connections among praxis, action research, and inter- and transdisciplinary methods make this type of research an example of convergence research or highly interdisciplinary study focused on addressing major existential challenges such as climate change and social transformation. Arctic sustainability research is a cutting-edge research area that expands the envelope of community research and defines the collaborative, post-disciplinary science of the future.

References cited

Adger, W. N. (2010). Social capital, collective action, and adaptation to climate change. In M. Voss (Ed.), *Der Klimawandel: Sozialwissenschaftliche Perspektiven* (pp. 327–345). Wiesbaden: VS Verlag für Sozialwissenschaften.

AHDR. (2015). *Arctic human development report: Regional processes and global linkages*. Copenhagen: Nordisk Ministerråd.

Alessa, L., Kliskey, A., & Williams, P. (2010). Forgetting freshwater: Technology, values, and distancing in remote Arctic communities. *Society & Natural Resources, 23*(3), 254–268.

Allen, C. R., Angeler, D. G., Cumming, G. S., Folke, C., Twidwell, D., Uden, D. R., & Bennett, J. (2016). Quantifying spatial resilience. *Journal of Applied Ecology, 53*(3), 625–635.

Allen, C. R., Angeler, D. G., Garmestani, A. S., Gunderson, L. H., & Holling, C. S. (2014). Panarchy: Theory and application. *Ecosystems, 17*(4), 578–589.

AMAP. (2017a). *Adaptation Actions for a Changing Arctic (AACA)—Bering/Chukchi/Beaufort region overview report*. Oslo, Norway: AMAP.

AMAP. (2017b). Adaptation actions for a changing Arctic: Perspectives from the Barents Area.

AMAP. (2018). *Adaptation actions for a changing Arctic: Perspectives from the Baffin Bay/Davis Strait Region.* Oslo, Norway: Arctic Monitoring and Assessment Programme (AMAP).

Armitage, D. (2005). Adaptive capacity and community-based natural resource management. *Environmental Management, 35*(6), 703–715.

Armitage, D., Berkes, F., Dale, A., Kocho-Schellenberg, E., & Patton, E. (2011). Co-management and the co-production of knowledge: Learning to adapt in Canada's Arctic. *Global Environmental Change, 21*(3), 995–1004.

ARR. (2016). *Arctic resilience report* (M. Carson & G. Peterson, Eds.). Stockholm: Arctic Council, Stockholm Environment Institute and Stockholm Resilience Centre.

Balbi, S., & Giupponi, C. (2010). Agent-based modelling of socio-ecosystems: A methodology for the analysis of adaptation to climate change. *International Journal of Agent Technologies and Systems, 2*(4), 17–38.

Bali, A., & Kofinas, G. (2014). Voices of the Caribou people: A participatory videography method to document and share local knowledge from the North American human-Rangifer systems. *Ecology and Society, 19*(2), (16).

Bell, S., & Morse, S. (2001). Breaking through the glass ceiling: Who really cares about sustainability indicators? *Local Environment, 6*(3), 291–309.

Berkes, F., & Armitage, D. (2010). Co-management institutions, knowledge, and learning: Adapting to change in the Arctic. *Études/Inuit/Studies, 34*(1), 109–131.

Berkes, F., & Berkes, M. K. (2009). Ecological complexity, fuzzy logic, and holism in indigenous knowledge. *Futures, 41*(1), 6–12.

Berkes, F., & Ross, H. (2013). Community resilience: Toward an integrated approach. *Society & Natural Resources, 26*(1), 5–20.

Berkes, F., & Ross, H. (2016). Panarchy and community resilience: Sustainability science and policy implications. *Environmental Science & Policy, 61*, 185–193.

Berman, M., Nicolson, C., Kofinas, G., Tetlichi, J., & Martin, S. (2004). Adaptation and sustainability in a small arctic community: Results of an agent-based similation model. *Arctic, 57*(4), 401–414.

Blair, B. (2017). *Toward Arctic transformations and sustainability: Modeling risks and resilience across scales of government* (Ph.D.). University of Alaska Fairbanks.

Bodin, Ö., & Crona, B. I. (2009). The role of social networks in natural resource governance: What relational patterns make a difference? *Global Environmental Change, 19*(3), 366–374.

Borgatti, S. P., Mehra, A., Brass, D. J., & Labianca, G. (2009). Network analysis in the social sciences. *Science, 323*(5916), 892–895.

Brubaker, M., Berner, J., & Tcheripanoff, M. (2013). LEO, the local environmental observer network: A community-based system for surveillance of climate, environment, and health events. *Circumpolar Health Supplements, 72*, 513–514.

Brunet, N. D., Hickey, G. M., & Humphries, M. M. (2014). The evolution of local participation and the mode of knowledge production in Arctic research. *Ecology and Society, 19*(2).

BurnSilver, S., & Magdanz, J. (2019). Heterogeneity in mixed economies. *Hunter Gatherer Research, 3*(4), 601–633.

BurnSilver, S., Magdanz, J., Stotts, R., Berman, M., & Kofinas, G. (2016). Are mixed economies persistent or transitional? Evidence using social networks from Arctic Alaska. *American Anthropologist, 118*(1), 121–129.

Canada, G. o. (2008). Implementation of the 2030 agenda for sustainable development.

Carson, M., & Sommerkorn, M. (2017). A resilience approach to adaptation actions. In *Chapter 8 of adaptation actions for a changing Arctic: Perspectives from the Barents area* (pp. 195–218). Oslo, Norway: Arctic Monitoring and Assessment Programme (AMAP).

Chandler, D. (2013). Resilience and the autotelic subject: Toward a critique of the societalization of security. *International Political Sociology, 7*(2), 210–226.

Chapin, F. S. III, Corrine, K., Brinkman, T. J., Bronen, R., & Cochran, P. (2016). Community-empowered adaptation for self-reliance. *Science Direct,* NA(NA).

Chapin, F. S. III, Kofinas, G. P., & Folke, C. (Eds.). (2009). *Principles of ecosystem stewardship: Resilience-based natural resource management in a changing world*. New York, NY: Springer-Verlag.

Chapin, F. S. III., Stuart, F., Kofinas, G., & Folke, C. (2009). A framework for understanding change. In I. Chapin, F. Stuart, G. Kofinas, & C. Folke (Eds.), *Principles of ecosystem stewardship: Resilience-based management in a changing world* (pp. 3–28). New York, NY: Springer-Verlag.

Clark, W. C. (2007). Sustainability science: A room of its own. *Proceedings of the National Academy of Sciences, 104*(6), 1737–1738.

Cole, M. A. (1999). Limits to growth, sustainable development and environmental Kuznets curves: An examination of the environmental impact of economic development. *Sustainable Development, 7*(2), 87–97.

Collings, P. (2011). Economic strategies, community, and food networks in Ulukhaktok, Northwest territories, Canada. *Arctic, 64*(2), 2007–2219.

DeFries, R., & Nagendra, H. (2017). Ecosystem management as a wicked problem. *Science, 356*(6335), 265–270.

Dombrowski, K., Channell, E., Khan, B., Moses, J., & Misshula, E. (2013). Out on the land: Income, subsistence activities, and food sharing networks in Nain, Labrador. *Journal of Anthropology, 2013*, 11.

Eamer, J. (2004). Keep it simple and be relevant: The first nine years of the Arctic borderlands ecological knowledge co-op. In W. V. Reid, F. Berkes, T. Wilbanks, & D. Capistrano (Eds.), *Bridging scales and knowledge systems: Concepts and applications in ecosystem assessment* (pp. 185–206). Washington, DC: Island Press.

Ernst, K. M., & Riemsdijk, M. V. (2013). Climate change scenario planning in Alaska's national parks: Stakeholder involvement in the decision-making process. *Applied Geography, 45*, 22–28.

Euskirchen, E. S., Timm, K., Breen, A. L., Gray, S., Rupp, T. S., Martin, P., .. McGuire, A. D. (In Press). Co-producing knowledge: The integrated ecosystem model for resource management in Arctic Alaska. *Frontiers in Ecology and the Environment*.

Fehling, M., Nelson, B. D., & Venkatapuram, S. (2013). Limitations of the millennium development goals: A literature review. *Global Public Health, 8*(10), 1109–1122.

Flynn, M., Ford, J. D., Pearce, T., & Harper, S. L. (2018). Participatory scenario planning and climate change impacts, adaptation and vulnerability research in the Arctic. *Environmental Science & Policy, 79*, 45–53.

Freeman, M., & Carbyn, L. (Eds.). (1988). *Traditional knowledge and renewable resource management in northern regions.* Edmonton: Boreal Institute for Northern Studies, University of Alberta.

Frodeman, R. (2011). Interdisciplinary research and academic sustainability: Managing knowledge in an age of accountability. *Environmental Conservation, 38*(02), 105–112.

Gad, U. P., & Strandsbjerg, J. (Eds.). (2018). *The politics of sustainability in the Arctic: Reconfiguring identity, space, and time.* New York, NY: Routledge.

Garmestani, A. S., & Cabezas, C. R. A. H. (2008). Panarchy, adaptive management and governance: Policy options for building resilience. *Nebraska Law Review, 87.*

Griggs, D., Stafford-Smith, M., Gaffney, O., Rockström, J., Öhman, M. C., Shyamsundar, P., .. Noble, I. (2013). Sustainable development goals for people and planet. *Nature, 495*, 305.

Gunderson, L. H., & Holling, C. S. (Eds.). (2002). *Panarchy: Understanding transformations in human and natural systems.* Washington, DC: Island Press.

Halonen, M., Persson, Å., Sepponen, S., Siebert, C. K., Bröckl, M., Vaahtera, A., .. Isokangas, A. (2017). *Sustainable development action—the Nordic way: Implementation of the global 2030 agenda for sustainable development in Nordic cooperation.* Copenhagen: Nordic Council of Ministers.

Hamilton, L. C., & Lammers, R. B. (2011). Linking pan-Arctic human and physical data. *Polar Geography, 34*(1–2), 107–123.

Holling, C. S. (2001). Understanding the complexity of economic, ecological, and social systems. *Ecosystems, 4*(5), 390–405.

Holm, P., Goodsite, M. E., Cloetingh, S., Agnoletti, M., Moldan, B., Lang, D. J., et al. (2013). Collaboration between the natural, social and human sciences. *Global Change Research Environmental Science & Policy, 28*, 25–35.

Huntington, H. P. (2011). The local perspective. *Nature, 478*, 182.

IASSA. (2019). *Research principles.* Retrieved 2019, from www.iassa.org.

Keskitalo, E. C. H., & Kulyasova, A. A. (2009). The role of governance in community adaptation to climate change. *Polar Research, 28*(1), 60–70.

Klein, J. T. (2004). Prospects for transdisciplinarity. *Futures, 36*(4), 515–526.

Kofinas, G., Abdelrahim, S., Carson, M., Chapin, F. S. III, Clement, J., Fresco, N., .. Veazey, A. (2016a). Building resilience in the Arctic: From theory to practice. In M. Carson & G. Peterson (Eds.), *Arctic resilience report.* Stockholm: Arctic Council/Stockholm Environment Institute and Stockholm Resilience Centre.

Kofinas, G., Aklavik, A. Village, Crow, O., & McPherson, F. (2002). Community contributions to ecological monitoring: Knowledge co-production in the U.S.-Canada Arctic borderlands. In I. Krupnik & D. Jolly (Eds.), *The earth is faster now: Indigenous observations of Arctic environmental change* (pp. 54–91). Fairbanks: ARCUS.

Kofinas, G., BurnSilver, S., Magdanz, J., Stotts, R., & Okada, M. (2016b). Subsistence sharing networks and cooperation: Kaktovik, wainwright, and venetie Alaska. University of Alaska Fairbanks, BOEM Project Report Number 2015–023DOI:500

Kofinas, G., Clark, D., Hovelsrud, G. K., Alessa, L., Amundsen, H., Berman, M., ... Olsen, J. (2013). Adaptive and transformative capacity. In A. Council (Ed.), *Arctic resilience interim report 2013* (pp. 71–91). Stockholm: Stockholm Environment Institute and Stockholm Resilience Centre.

Kruse, J. A. (1991). Alaska Iñupiat subsistence and wage employment patterns: Understanding individual choice. *Human Organization, 50*(4), 317–326.

Kruse, J. A., & White, R. G. (1995). Suistainability of Arctic communities: Interactions between global changes, public policies, and ecological processes, NSF—Proposal.

Kruse, J. A., White, R. G., Epstein, H. E., Archie, B., Berman, M., Braund, S. R., . . Eamer, J. (2004). Modeling sustainability of Arctic communities: An interdisciplinary collaboration of researchers and local knowledge holders. *Ecosystems, 7*(8), 815–828.

Kumpula, T., Pajunen, A., Kaarlejärvi, E., Forbes, B. C., & Stammler, F. (2011). Land use and land cover change in Arctic Russia: Ecological and social implications of industrial development. *Global Environmental Change, 21*(2), 550–562.

Lang, D. J., Wiek, A., Bergmann, M., Stauffacher, M., Martens, P., Moll, P., . . Thomas, C. J. (2012). Transdisciplinary research in sustainability science: Practice, principles, and challenges. *Sustainability Science, 7*(1), 25–43.

Larsen, J. N., & Fondahl, G. (2015). *Arctic human development report: Regional processes and global linkages* (p. 507). Copenhagen, Denmark: Norden.

Larsen, J. N., Schweitzer, P., & Fondahl, G. (Eds.). (2010). *Arctic social indicators—a follow up to the Arctic human development report.* Copenhagen: Nordic Council of Ministers.

Larsen, J. N., Schweitzer, P., & Petrov, A. (2015). *Arctic social indicators: ASI II: Implementation.* Copenhagen: Nordic Council of Ministers.

Lather, P. (1986). Research as praxis. *Harvard Educational Review, 56*(3), 257–278.

Ledford, H. (2015). Team Science: Interdisciplinarity has become all the rage as scientists tackle society's biggest problems. But there is still strong resistance to crossing borders. *Nature, 525*(7569), 308–311.

Levin, S. A. (1999). Towards a science of ecological management. *Conservation Ecology, 3*(2), 6.

Lin, N. (2002). Building a network theory of social capital. In N. Lin, K. S. Cook, & R. S. Bur (Eds.), *Social capital: Theory and research* (pp. 3–30). New York, NY: Gruyter.

Lubell, M., Robins, G., & Wang, P. (2014). Network structure and institutional complexity in an ecology of water management games. *Ecology and Society, 19*(4).

Magdanz, J. S., Utermohle, C. J., & Wolfe, R. J. (2002). *The production and distribution of wild food in wales and Deering, Alaska.* Technical Paper 259. Alaska Department of Fish and Game, Alaska.

Mahoney, J. (2000). Path dependence in historical sociology. *Theory and Society, 29*, 507–548.

Meek, C. L. (2013). Forms of collaboration and social fit in wildlife management: A comparison of policy networks in Alaska. *Global Environmental Change, 23*(1), 217–228.

Moen, J., & Keskitalo, E. C. H. (2010). Interlocking panarchies in multi-use boreal forests in Sweden. *Ecology and Society, 15*(3), 17.

Moon, T. A., Overeem, I., Druckenmiller, M., Holland, M., Huntington, H., Kling, G., . . Wong, G. (2019). The expanding footprint of rapid Arctic change. *Earth's Future, 7*(3), 212–218.

Nadasdy, P. (1999). The politics of TEK: Power and the 'Integration' of knowledge. *Arctic Anthropology, 36*(1–2), 1–18.

Natcher, D. C. (2015). Social capital and the vulnerability of aboriginal food systems in Canada. *Human Organization, 74*(3), 230–242.

Neff, T. (2009). Connecting science and policy to combat climate change.

Nel, J. L., Roux, D. J., Driver, A., Hill, L., Maherry, A. C., Snaddon, K., . . Reyers, B. (2016). Knowledge co-production and boundary work to promote implementation of conservation plans. *Conservation Biology, 30*(1), 176–188.

Neumann, R. P. (2009). Political ecology: Theorizing scale. *Progress in Human Geography, 33*(3), 398–406.

Nicholson, C. R., Starfield, A. M., Kruse, J., & Kofinas, G. P. (2002). Ten heuristics for interdisciplinary modeling projects. *Ecosystems, 5*(6), 376–3384.

Nilsson, A. E., Bay-Larsen, I., Carlsen, H., van Oort, B., Bjørkan, M., Jylhä, K., . . van der Watt, L.-M. (2017). Towards extended shared socioeconomic pathways: A combined participatory bottom-up and top-down methodology with results from the Barents region. *Global Environmental Change, 45*, 124–132.

Nilsson, A. E., Carson, M., Cost, D. S., Forbes, B. C., Haavisto, R., Karlsdottir, A., . . Pelyasov, A. (2019). Towards improved participatory scenario methodologies in the Arctic. *Polar Geography*, 1–15.

North, D. C. (1990). *Institutions, institutional change and economic performance.* New York, NY: Cambridge University Press.

Olsson, L., Jerneck, A., Thoren, H., Persson, J., & O'Byrne, D. (2015). Why resilience is unappealing to social science: Theoretical and empirical investigations of the scientific use of resilience. *Science Advances, 1*(4), e1400217.

Orttung, R., & Laruelle, M. (Eds.). (2017). *Urban sustainability in the Arctic. Visions, contexts, challenges.* Washington, DC: IERES.

Ostrom, E. (2005). *Understanding institutional diversity.* Princeton: Princeton University Press.

Ozkan, U. R., & Schott, S. (2013). Sustainable development and capabilities for the polar region. *Social Indicators Research, 114*(3), 1259–1283.

Petrov, A. N. (2018). Summary and updates: IASSA participation in the Arctic council meetings in fall 2018. *Northern Notes, 50*, 5.

Petrov, A. N., BurnSilver, S., Chapin, F. S., Fondahl, G., Graybill, J., Keil, K., . . Schweitzer, P. (2016). Arctic sustainability research: Toward a new agenda. *Polar Geography, 39*(3), 165–178.

Petrov, A. N., BurnSilver, S., Chapin, F. S. III, Fondahl, G., Graybil, J. K., Keil, K., . . Schweitzer, P. (2017). *Arctic sustainability research: Past, present and future*. Routledge.

Planque, B., Mullon, C., Arneberg, P., Eide, A., Fromentin, J.-M., Heymans, J. J., . . Thorvik, T. (2019). A participatory scenario method to explore the future of marine social-ecological systems. *Fish and Fisheries, 20*(3), 434–451.

Pohl, C., Rist, S., Zimmermann, A., Fry, P., Gurung, G. S., Schneider, F., . . Wiesmann, U. (2010). Researchers' roles in knowledge co-production: Experience from sustainability research in Kenya, Switzerland, Bolivia and Nepal. *Science and Public Policy, 37*(4), 267–281.

Putnam, R. D. (2007). E pluribus Unum: Diversity and community in the twenty-first century the 2006 Johan Skytte prize lecture. *Scandinavian Political Studies, 30*(2), 137–174.

Ready, E., & Power, E. A. (2018). Why wage earners hunt: Food sharing, social structure, and influence in an Arctic mixed economy. *Current Anthropology, 59*(1), 74–97.

Reid, J. (2013). Interrogating the neoliberal biopolitics of the sustainable development-resilience nexus. *International Political Sociology, 7*(4), 353–367.

Reid, W. V., Berkes, F., Wilbanks, T. J., & Capistrano, D. (Eds.). (2006). *Bridging scales and knowledge systems*. Washington, DC: Island Press.

Rittel, H. T., & Webber, M. M. (1973). 2.3 planning problems are wicked. *Polity, 4*, 155–169.

Robinson, G. M., & Carson, D. A. (2016). Resilient communities: Transitions, pathways and resourcefulness. *The Geographical Journal, 182*(2), 114–122.

Rotmans, J., & Rotmans, D. S. (Eds.). (2003). *Scaling in integrated assessment*. Lisse: Swets and Zeitlinger.

Sachs, J. D. (2012). From millennium development goals to sustainable development goals. *The Lancet, 379*(9832), 2206–2211.

Scheffer, M., Bolhuis, J. E., Borsboom, D., Buchman, T. G., Gijzel, S. M. W., Goulson, D., . . Olde Rikkert, M. G. M. (2018). Quantifying resilience of humans and other animals. *Proceedings of the National Academy of Sciences, 115*(47), 11883–11890.

Stokes, D. E. (1997). *Pasteur's quadrant*. Washington, DC: Brookings Institution Press.

Suter, L., Schaffner, C., Giddings, C., Orttung, R., & Streletskiy, D. D. (2017). Developing metrics to guide sustainable development of Arctic cities: Progress & challenges. In *Arctic yearbook*. Akureyri, Iceland: Northern Research Forum.

Tengö, M., Brondizio, E. S., Elmqvist, T., Malmer, P., & Spierenburg, M. (2014). Connecting diverse knowledge systems for enhanced ecosystem governance: The multiple evidence base approach. *AMBIO, 43*(5), 579–591.

Thornton, T. F. (2001). Subsistence in northern communities: Lessons from Alaska. *Northern Review, 23*, 82–102.

UN. (2007). *The millennium development goals report 2007.* New York, NY: United Nations, Retrieved August 19, 2019.

UN. (2015). *Transforming our world: The 2030 agenda for sustainable development.* New York, NY: United Nations.

Valentin, A., & Spangenberg, J. H. (2000). A guide to community sustainability indicators. *Environmental Impact Assessment Review, 20*(3), 381–392.

van der Hel, S. (2016). New science for global sustainability? The institutionalisation of knowledge co-production in Future Earth. *Environmental Science & Policy, 61,* 165–175.

Vargas-Moreno, J. C., Fradkin, B., Emperador, S., & Lee, O. (Eds.). (2016). *Project summary: Prioritizing science needs through participatory scenarios for energy and resource development on the north slope and adjacent seas.* Boston, MA: GeoAdaptive, LLC.

Vlasova, T., & Volkov, S. (2016). Towards transdisciplinarity in Arctic sustainability knowledge co-production: Socially-Oriented observations as a participatory integrated activity. *Polar Science, 10*(3), 425–432.

Voorberg, W. H., Bekkers, V. J. J. M., & Tummers, L. G. (2015). A systematic review of co-creation and co-production: Embarking on the social innovation journey. *Public Management Review, 17*(9), 1333–1357.

Votrin, V. (2006). *Measuring sustainability in the Russian Arctic: An interdisciplinary study* (Ph.D. dissertation). University of Brussels.

Walker, B., & Salt, D. (2006). *Resilient thinking: Sustaining ecosystems and people in a changing world.* Washington, DC: Island Press.

Weitz, N., Persson, Å., Nilsson, M., & Tenggren, S. (2015). *Sustainable development goals for Sweden: Insights on setting a national agenda.* Working Paper 2015–10. Stockholm Environment Institute, Stockholm.

Wenzel, G. W. (1995). Ningiqtuq: Resource sharing and generalized reciprocity in Clyde River, Nunavut. *Arctic Anthropology, 32*(2), 43–60.

Wheeler, P., & Thornton, T. (2005). Subsistence research in Alaska: A thirty-year retrospective. *Alaska Journal of Anthropology, 3*(1), 69–103.

Wiek, A., Farioli, F., Fukushi, K., & Yarime, M. (2012). Sustainability science: Bridging the gap between science and society. *Sustainability Science, 7*(1), 1–4.

Wolfe, R. J., Scott, C. L., Simeone, W. E., Utermohle, C. J., & Pete, M. C. (2009). *The "Super-Household" in Alaska native subsistence economies.* Washington, DC: National Science Foundation.

Wu, J., & Wu, T. (2012). Sustainability indicators: An overview. In C. N. Madu & C.-H. Kuei (Eds.), *Handbook of sustainability management* (pp. 65–86). London: World Scientific Publishing.

Young, O. R. (2002). Institutional interplay: The environmental consequences of cross-scale interactions. In E. Ostrom, T. Deitz, N. Dolsak, P. C. Stern, S. Stonich, & E. U. Weber (Eds.), *The drama of the commons.* Washington, DC: National Academy Press.

Young, O. R. (2012). Arctic tipping points: Governance in turbulent times. *AMBIO, 41*(1), 75–84.

Young, O. R. (2016). *On environmental governance: Sustainability, efficiency, and equity.* New York: Routledge.

Zijp, M. C., Posthuma, L., Wintersen, A., Devilee, J., & Swartjes, F. A. (2016). Definition and use of solution-focused sustainability assessment: A novel approach to generate, explore and decide on sustainable solutions for wicked problems. *Environment International, 91*, 319–331.

Ziker, J., & Schnegg, M. (2005). Food sharing at meals. *Human Nature, 16*(2), 178–210.

Afterword

In anticipation of *Arctic Sustainability, Community, and Environment: A Synthesis of Knowledge II*

Andrey N. Petrov, Jessica K. Graybill, Tatiana Degai, Aileen A. Espíritu, Diane Hirshberg, and Tatiana Vlasova

Volume I of this two-part book series covered several key topics in current Arctic sustainability research. Discussing theoretical and methodological foundations of the emerging Arctic sustainability science and highlighting state-of-the-art research in economic and cultural sustainability, sustainable governance, and resource management, Volume I provides a foundation for further discussions on various aspects of sustainability and sustainable development research in the Arctic. Volume II will focus on specific and applied elements of Arctic sustainability, including urban and environmental sustainability, Indigenous perspectives on sustainability in Arctic communities, and monitoring for sustainability.

"Sustainable Environments" will address the state of the science about sustainable Arctic environments, including theoretical and methodological approaches and case studies.

"Sustainable Cities" will address Arctic cities and associated movements and ideas about sustainability for them. Recent advancements, knowledge gaps, and future priorities will be addressed in addition to theoretical and methodological approaches.

"Indigenous Perspectives on Sustainability" will address Indigenous visions of sustainability and sustainable development and Indigenous knowledge engagement in sustainability research. Core principles of sustainability from Indigenous perspectives will outlines challenges to indigenous sustainability and explain the role of Indigenous knowledge and scholarship in sustainable development solutions in the Arctic.

"Monitoring for Sustainability" will address the principle of and methods for monitoring sustainable development, including discussion of sustainability as a process and an outcome. Major monitoring systems, recent advancements, knowledge gaps, and future priorities will be discussed.

"Sustainable Development Goals and the Arctic" will examine the application of the United Nations Sustainable Development Goals to the Arctic and the involvement of Arctic scholars, policymakers, and communities in developing and implementing this agenda.

"State of Knowledge, Gaps, and a Future Agenda" will provide a synthesis of the findings from Volumes I and II and will outline the contours of the modern Arctic sustainability research, including advancements, knowledge gaps, and future priorities.

Index

Note: Page numbers in *italics* indicate figures and page numbers in **bold** indicate tables on the corresponding pages.